Weather Eye

TO ANNE
SINE QUA NON

Weather Eye

BRENDAN McWILLIAMS

THE LILLIPUT PRESS
MCMXCIV

First published in 1994 by
THE LILLIPUT PRESS LTD
4 Rosemount Terrace, Arbour Hill,
Dublin 7, Ireland.

A CIP record for this
title is available from
The British Library.

ISBN 1 874675 38 4

Cover design by Howard Noyes
Set in 10 on 12.5 Adobe Garamond by
mermaid turbulence
Printed in Dublin by Betaprint

Contents

Acknowledgments

I am grateful to successive Directors of the Meteorological Service for their support and encouragement over the years; to Lisa Sheilds, the Service's Librarian, for her unstinting and never-failing efforts to provide precisely whatever reference might be needed; and to my colleagues in the Meteorological Service for their help in answering my silly questions, and their frequent tolerance at finding what appeared to be a casual conversation over coffee reproduced in *The Irish Times* a few days later.

The editorial team at *The Irish Times* deserve my thanks for their eternal vigilance in weeding out the worst excesses of the ever-present gremlins. I am also in debt to the hundreds of readers who have written to me over the years with comments, suggestions and details of their own experiences in the world of meteorology; I have replied to most of them – but alas not all, and I hope that those whose overtures have met with silence will rest assured that their information has been read, appreciated, and very often used. At the very least it is reassuring to learn that someone actually reads the words one writes.

Most important of all, I am grateful to my wife Anne for her advice, research, encouragement and solid practical help in producing the daily column down the years. She has cheerfully tolerated the solitude of my nightly sojourn to the keyboard in the interests of meteorology and art, and without her this book would not exist. And finally I thank Stephen and Laurie, whose irrelevant and irreverent advice and welcome interruptions have helped to make the journey tolerable.

Foreword

The very first 'Weather Eye' appeared in *The Irish Times* on 9 August 1988. A month or so previously I had submitted to that paper a series of four feature-length articles on meteorology; I assume the reaction of the readers must have been benign, because shortly afterwards I was asked if I would consider the tyranny of a daily column – and being younger and less wise than I am now, I readily agreed. Since then, nearly two thousand individual pieces have appeared, and although it is not always easy to maintain the pace, it is a discipline I have come to quite enjoy.

As regular readers of the column will be well aware, the subject matter of 'Weather Eye' strays frequently from the narrow path of meteorology regarded purely as a science. I have been conscious that if the reader, and indeed I myself, merely wanted to find out how the weather works, it is only necessary to consult a standard text-book on the subject. But behind the science lie the stories of the people who developed it, of the mistakes they made along the way, and of the plausible misapprehensions in which they found their inspiration. These, even more than the wonders of the atmosphere and the miracles achieved by modern technology, lend vitality and fascination to the subject.

These stories are not difficult to find. Anyone assisted by the 'six honest serving men' described by Kipling is faced with a myriad of unanswered questions on any weather topic that might come to mind. The six, if you remember, were 'What and Why and When, and How and Where and Who': they, combined with a certain familiarity with the classics which perhaps betokens a youth that was not sufficiently misspent, have helped to make 'Weather Eye' whatever it may be to you.

Almost since the very first article appeared, there have been frequent requests that an anthology be published in more permanent form. I have been slow to rise to the occasion, and have found it difficult to choose a selection that might encapsulate the flavour of the daily column. But here they are at last – over a hundred 'Weather Eyes' that I hope may be enjoyed.

BRENDAN MCWILLIAMS

June 1994

Quelqu'un pourrait dire de moi que j'ai seulement fait ici un amas de fleurs étrangères, n'y ayant fourni du mien que le filet à les lier.

Montaigne, *Essais*, III, xii

It could be said of me that in this book I have merely made up a bunch of other people's flowers, providing from myself only the string that ties them all together.

I *Searching for the Secrets*

A Talent to Peruse

The forecasters of ancient Rome occupied a position of great power and influence. Unlike today's practitioners, they were not obliged to confine themselves to predictions of the weather. The Roman augurs had a wide remit, and their pronouncements on the likely course of future events were awaited with eager anticipation at the start of any important enterprise.

It was the duty of the augurs to observe the signs – or *auspices* – which were sent by the gods to indicate their approval or otherwise of any proposed undertakings. The auspices took many forms. Signs from the birds – relating to their pattern and direction of flight, to the sounds they made, and to the ways in which they took their food – were of particular significance.

In general, signs from the right-hand side were considered to be good, while those which manifested themselves from the left were unlucky or sinister. Indeed they were literally so: *sinister* is the Latin word for 'left'. But the most trusted sources of information for the augurs were the entrails of sacrificial animals. The liver was found to be particularly reliable in this regard, because of the subtle variations to be found in its size and shape, and in its colour and the pattern of its veins.

The operation was taken very seriously. Before taking the auspices the augur marked out the *templum*, or consecrated space, within which his observations were intended to be made. Anything outside the templum did not count; within its limits, the augur pitched his tent, asked the gods for signs, and waited for his answer.

Since magistrates were legally bound to take appropriate action on the advice of an augur, the office could be used by unscrupulous practitioners for personal political purposes. An unfavourable report could be used to obtain the postponement of unwanted meetings of the Senate, or to cancel the results of an election whose outcome might prove to be somewhat inconvenient. In 59 BC, for example, the augur Bibulus succeeded in holding up the entire legislative programme of Julius Caesar by merely, as he put it, 'watching the heavens'.

For these sensitive reasons, the office of augur was bestowed only on persons of the most distinguished merit. This tradition of excellence has continued for those required to gaze into the future nowadays but the power, the influence, the flamboyant trappings of office, and indeed the talent for omniscience, have long since disappeared. *O tempora, O mores!*

4

An Early Enthusiast

'Those whom the gods love die young' according to the Greek playwright Menander. Perhaps that was the case with William Molyneux, who died on 11 October 1698, at the early age of forty-two. But despite his somewhat premature demise, William Molyneux has achieved a lasting place in climatological history; he is credited with being Ireland's very first scientific weather observer.

Molyneux was educated at Trinity College, Dublin, and later studied law in London. His interests throughout his life were wide, and he was no stranger to political controversy. Indeed only a few months before he died, he wrote 'The Case of Ireland's being bound by Acts of Parliament in England Stated', a tract which attracted sufficient attention for it to be condemned by the London parliament in June of that year for being 'of dangerous tendency to the Crown and to the people of England.'

Molyneux lived in an era of rapid development in the field of scientific instrumentation, and he quickly realized the potential of these new instruments for gaining an insight into the behaviour of the atmosphere. He was discouraged, however, by his difficulty in acquiring them: 'I am living in a kingdom barren of all things', he lamented in 1681, 'but especially of ingenious artificers; I am wholly destitute of instruments on which I can rely.'

But the situation did improve. In March of 1684 Molyneux was able to begin a 'Weather Register', which for the first time in Ireland included readings of barometric pressure. By June 2nd of that year he had compiled enough material to present a paper to the Dublin Society on 'The Observations of the Weather for the Month of May, with the Winds and the Heights of the Mercury in the Baroscope.' He sent a copy of his May Register to Oxford University where it remains to this day, preserved in the Bodleian Library.

In May of the following year, Molyneux handed over the exacting task of keeping weather records for Dublin to St George Ashe, later to become Provost of Trinity College, and Ashe maintained the continuity for another year or so. This series of observations, although it only lasted for the two-year period 1684–6 and only a small fragment of it still survives, is regarded as one of the most important milestones in the history of Irish meteorology.

From Men-o'-War to Bits of Paper

Long before the invention of instruments that could measure the speed of the wind accurately, people used to guess at it – and then describe it. But for hundreds of years such descriptions were purely subjective. Who would have thought, for example, that the pirate William Dampier, writing in 1697 and describing the wind as merely 'blowing exceeding hard', was in the middle of a full typhoon? At other times, however, gross exaggeration was the order of the day. Admiral Sir Francis Beaufort was the first to standardize the measurement of wind, and the scale which bears his name has survived since the beginning of the last century with changes of a mere cosmetic nature.

Beaufort was born in 1774 into a family of French Huguenot origin in Co. Louth, where his father, Dr Daniel Augustus Beaufort, was rector of the local church. At the tender age of fourteen, young Francis embarked on a naval career, his family having paid the not inconsiderable sum of 100 guineas for the privilege; he was taken aboard the good ship *Vansittart* at Gravesend on 20 March 1789, and in due course crowned a distinguished career by becoming Hydrographer to the Royal Navy and being made a knight.

It was in 1805 that Beaufort's scale of wind force was officially adopted. For the lower range of his thirteen-point scale, he took his cue from the descriptive terms traditionally used by sailors. Force 0 was a 'calm', Force 1 a 'light air', and Force 2 a 'slight breeze'. For the stronger winds, he realized that he had to define his scale in terms of some well-known yardstick, just as a standard measure might be used to determine the length or weight of another object. The criterion he chose was the full-rigged battleship or 'man-o'-war' of his day. He described the winds by the effect they might have on such a vessel – and in particular the amount of sail it could carry in high winds without getting into trouble.

A century later, in the early 1900s, there were no longer any men-o'-war by which to learn the wind speeds, and so the descriptions had to be revised. For maritime purposes, the winds were now defined in terms of their effect on the surface of the open sea. Force 8 or Gale Force, for example, represents winds averaging slightly over 40 miles per hour, and was described by Beaufort as a wind in which 'a well-conditioned man-o'-war might carry triple reef and courses'; the new Gale Force 8 resulted in 'moderately high waves, where ... the foam is blown in well marked streaks along the direction of the wind.'

A further change to the Beaufort Scale was necessary to cater for the vast majority of the population who, like W.S. Gilbert's Admiral, 'stick close to their desks and never go to sea'. They were accommodated by Sir George Simpson, who in 1906 related the Beaufort numbers to familiar homely things like loose bits of paper, umbrellas, trees and chimney pots. And this, in essence, is the Beaufort Scale of Wind Force that is still in use today.

Rear-Admiral Sir Francis Beaufort (1774–1857). (Painting by S. Pearce, 1851, National Maritime Museum, Greenwich)

Drains, Dykes and Weather-Maps

The inspection of drains appears to nurture creativity. Percy French, for example, before achieving more lasting fame as the composer of many of our best-loved Irish airs, began his career as the official inspector of drains for Co. Cavan. But far away, and a longer time ago, another member of the hydrological inspectorate carved out his own particular niche in history.

Heinrich Wilhelm Brandes was born on 30 July 1777, in the little German town of Ritzebuttel. Brandes spent the first ten years of his adult life as an Inspector of Dykes on the River Weser, and might, had his talents led in that direction, have progressed to write Teutonic gems like 'How are things in Ritzebuttel?' Instead, however, he became a meteorologist, and he is credited with drawing the very first weather-map.

With a growing reputation as a mathematician, Brandes was appointed Professor of Mathematics at the University of Breslau in 1811, and it was there that he developed his interest in meteorology. The science was still in its infancy. From 1600 onwards, the invention of many of the now familiar meteorological instruments made scientific weather observations possible for the first time. It was nearly 200 years later, however, before any serious attempt was made to obtain simultaneous readings from a large number of places. And even then, nothing very much was done with them – until Brandes came along.

The most successful observing network of the era was one of thirty-seven stations throughout Europe, organized during the 1780s by what was called the Meteorological Society of Mannheim. Although the project was in operation for only twelve years, it was a very valuable step forward; for the first time observations were carried out with standardized methods and with carefully calibrated instruments.

More than thirty years later, during the period from 1816 to 1821, Brandes decided that there was a great deal to be learned from this valuable series of data. He entered the observations for each day of the year 1783 on 365 individual maps, and then drew lines to indicate deviations of the atmospheric pressure from its normal value. From these charts he hoped it would be possible to determine 'the limit of the large rain cloud which lies over Germany and France in July...' His charts identified for the first time the depressions and anticyclones which are so familiar to us today, and thus laid the foundations of modern meteorology.

8

A Storm of Consequence

Shipwrecks are still a fact of life. The sea remains the scavenger it always was, a fickle opportunist, waiting with Pavlovian anticipation to subsume the unwary mariner into the oblivion of its murky depths. 'The sea', as Joseph Conrad said, 'has no generosity.'

Wrecks, however, are not as common as they used to be. Stronger and more sturdy ships, better communications between ship and shore, modern navigational aids, and above all, accurate and timely weather forecasts, have all deprived the sea of countless victims. But in the middle of the last century it had its share; for example over 200 vessels, large and small, were wrecked in the waters around these islands between 21 October and 2 November 1859. The best remembered is the *Royal Charter.*

On 25 October 1859 the steamship *Royal Charter* called at Cobh (then Queenstown) in Co. Cork. It had left Melbourne, Australia, just two months previously and was on its way to Liverpool with 430 passengers and crew, and a cargo of £500,000 in gold bullion. The stay in Cork was brief, and the vessel left Queenstown later that day on the final leg of its long voyage. At 3 a.m. on the twenty-sixth, a vicious storm drove the ship ashore near Moelfe on the north-east coast of the isle of Anglesea. Within five hours it had been dashed to pieces; a few of those on board were saved, but over 400 persons perished in the wreck.

Although in those days organized meteorology was in its infancy, it had reached a stage where regular weather observations were carried out at a few places, and these records have made it possible to reconstruct the progress of the Royal Charter storm. It seems to have developed west of the Azores, and by 9 a.m. on 25 October a depression of 965 millibars, or hectopascals as they are now called, had reached Brest in northern France. From there it moved towards Plymouth, passing over the Eddystone lighthouse, and headed northwards over Wales. At 6 a.m. on the twenty-sixth, it was seventy or eighty miles to the east of Anglesea – and the *Royal Charter.*

The tragedy is well remembered by meteorologists. It was as a result of the publicity given to this storm and the subsequent enquiry that the head of the 'Meteorological Department' of the British Navy, Vice-Admiral Robert FitzRoy, was charged with organizing a system of 'storm warnings' which were to be sent to threatened coastal areas.

FitzRoy organized a network of forty weather stations around the

Irish and British coastlines, which provided him with daily weather reports by electric telegraph. His forecasting methods were primitive by today's standards, although he was not one to oversimplify them for the benefit of the man in the street: 'The weather of our country', he told anyone who was interested, 'usually depends completely on the collision, combination, alternating predominance, or successive exchanges of parts of competing polar and equatorial countercurrents.' George Bernard Shaw – whose theory, if you remember, was that all professions are conspiracies against the laity – could reasonably claim a QED!

FitzRoy was a practical man at heart. He produced the required forecasts, and instituted the storm warnings which began in February of 1861. When gales were expected, warnings were telegraphed to forty ports and harbours around the country; within thirty minutes appropriate signals were prominently displayed on shore to relay the word to passing ships. The signals were of a semaphore type: a cone pointing upwards meant a northerly gale; a drum or cylinder warned of successive gales from many directions; and other patterns had meanings which quickly became standard and widely understood.

FitzRoy's storm warnings were noted with some scepticism by the general public, but were staunchly defended, and widely used and appreciated, by the shipping fraternity of the day. Although far from infallible, he was right often enough to be useful. Indeed four years later, on hearing of FitzRoy's death, the wife of an Aberdeen fisherman was heard to exclaim: 'Who will look after our men now?'

Robert FitzRoy (1805–65) as a young man.

The First of Many

At eight o'clock in the morning on 8 October 1860 the very first Irish 'real time' weather observation was transmitted from Valentia Island in Co. Kerry. Weather reports have been coming in a continuous stream from that part of the country ever since, albeit not from precisely the same spot.

The historic message was sent on the electric telegraph to FitzRoy in London for use in his newly organized system of storm warnings. The observation was performed by Mr R.J. Lecky, who at the time was manager of the telegraphic station on Valentia Island. Lecky continued with this valuable service on a daily basis for many years; he was made redundant only by the establishment of an official Observatory on the island on 15 June 1868.

The new Observatory was at first a very modest undertaking. It occupied a rented house on the narrow strait which separates Valentia Island from the rest of Kerry, and from there the routine flow of observations continued until March of 1892. It was in that year that Valentia Observatory moved across the sound to its present site on the mainland near the town of Cahirciveen, retaining for *auld lang syne* its traditional name – the name by which, somewhat confusingly, it is still known.

Its new home was Westwood House, theretofore the residence of one Captain Needham, the local agent of Trinity College which was at that time a very prominent landowner in the locality. Westwood was purchased for the not inconsiderable sum of £1400, and in the succeeding years was decked out with the impressive array of scientific instruments which was in due course to make Valentia Observatory one of the most important meteorological and geophysical observatories in all of Western Europe.

Valentia Observatory is today one of the Irish weather-observing stations whose hourly reports of current weather conditions are circulated around the globe; it performs upper-air measurements, using a *radiosonde* attached to a hydrogen-filled balloon to obtain values of pressure, temperature and humidity many miles above the earth; it monitors variations in the earth's magnetic field, and carries out precise measurements of radiation coming from the sun; the Observatory also operates a seismograph, which detects and records tiny vibrations which may have their origins thousands of miles away in earthquakes half-way around the world.

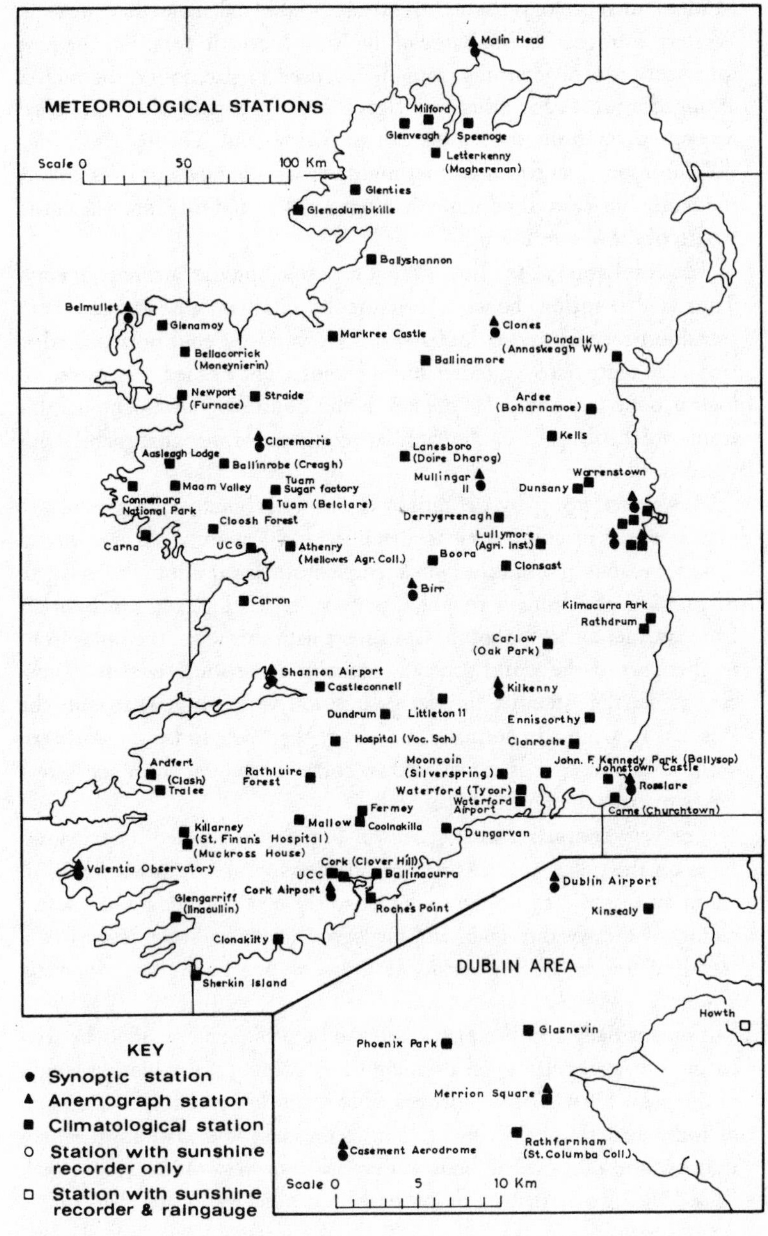

METEOROLOGICAL STATIONS
Scale 0 50 100 Km
Malin Head
Milford
Glenveagh
Speenoge
Letterkenny (Magherenan)
Glenties
Glencolumbkille
Ballyshannon
Belmullet
Glenamoy
Markree Castle
Clones
Dundalk (Annaskeagh WW)
Bellacorrick (Moneynierin)
Ballinamore
Newport (Furnace)
Straide
Ardee (Boharnamoe)
Kells
Claremorris
Lanesboro (Doire Dharog)
Aasleagh Lodge
Ballinrobe (Creagh)
Mullingar II
Warrenstown
Dunsany
Maam Valley
Tuam Sugar factory
Tuam (Belclare)
Connemara National Park
Derrygreenagh
Cloosh Forest
Lullymore (Agri. Inst.)
Carna
UCG
Athenry (Mellowes Agr. Coll.)
Boora
Clonsast
Birr
Carron
Kilmacurra Park
Rathdrum
Carlow (Oak Park)
Shannon Airport
Castleconnell
Kilkenny
Littleton 11
Dundrum
Enniscorthy
Hospital (Voc. Sch.)
Clonroche
Ardfert (Clash)
Tralee
Rathluirc Forest
Mooncoin (Silverspring)
John. F. Kennedy Park (Ballysop)
Johnstown Castle
Rosslare
Waterford (Tycor)
Waterford Airport
Carne (Churchtown)
Fermoy
Mallow
Coolnakilla
Dungarvan
Killarney (St. Finan's Hospital)
(Muckross House)
Valentia Observatory
Cork (Clover Hill)
UCC
Ballinacurra
Cork Airport
Roche's Point
Glengarriff (Ilnacullin)
Clonakilty
Sherkin Island
DUBLIN AREA
Dublin Airport
Kinsealy
Howth
Glasnevin
Phoenix Park
Merrion Square
Rathfarnham (St. Columba Coll.)
Casement Aerodrome
Scale 0 5 10 Km
KEY
Synoptic station
Anemograph station
Climatological station
Station with sunshine recorder only
Station with sunshine recorder & raingauge

The Winds at War

Modern meteorology has as its metaphor the stalemate that arose in Western Europe after the Battle of the River Marne in 1914. For the next four years two huge armies, roughly balanced in size, found themselves arranged against each other in a zigzag line of trenches which stretched for 300 miles from the North Sea to Switzerland. During that time, despite minor incursions costing hundreds of thousands of lives apiece, the battle line swayed no more than ten miles to and fro along the entire length of the Western Front.

Meteorologists at the time were just developing the 'air mass' theory. They saw an analogy between the current military impasse and the sharp transition zone in mid-latitudes which separates the cold polar easterlies from the temperate westerlies further south. They called this meteorological boundary the *polar front*; it is the transitory movements of this front which bring to us the familiar sequences of our changeable Irish weather.

A weather front, by definition, is simply a boundary between two masses of air of contrasting temperature and humidity. As one would expect, bearing in mind the global temperature distribution, the 'normal' orientation of a front is roughly east-west, separating cold northern air from warmer air to the south. Consistent with this ideal, the polar front in this part of the world generally stretches from south-west to north-east across the Atlantic. Its average position varies somewhat with the seasons. In winter it normally runs from the West Indies to Portugal, while in summer it is much farther north, stretching from the Great Lakes towards the north of Scotland.

For very complex reasons, areas of low pressure called 'depressions' form on the polar front. As depressions develop, they cause a wedge of warm air to project into the colder air to the north – the familiar 'warm sector' of the weather-map, and analogous in the military context to a transitory incursion into enemy territory by one of the two opposing armies.

On our daily weather-map a series of depressions can often be seen strung out at regular intervals along the continuous, undulating length of the polar front. The weather sequence seems like an endless succession of individual rain-belts, but it is in fact usually the *same* front which returns time after time to bestow upon us its unwelcome attention, each episode being a battle won or lost in the perennial war of the elements.

II *The Elements Explained*

The Dividing Lines

As a front moves along the landscape, it brings an area which has been under the influence of one kind of air into a new regime with quite different characteristics. In general, the air on one side of a front is warm and humid, and that on the other side cooler and relatively dry. The boundary between the two is surprisingly well-defined, and as the two air masses sweep along the sudden change in temperature and humidity is very obvious from instrumental records as the front passes a particular spot. If the passage of a front results in the existing air mass being replaced by a warmer one, the front in question is a *warm* front; if the temperature afterwards is lower than before, then a *cold* front has passed.

The barbs drawn on a front indicate what kind of front it is. A warm front is identified by semicircular barbs; a cold front by triangular ones, and the barbs point in the direction towards which the front is expected to move. If the weather-map is in colour, each front can be even more clearly distinguished, since warm fronts are red, and cold fronts blue.

If you look closely at a weather-chart, you may notice that what is designated a warm front for part of its length, is a cold front for the remainder. This is not an attempt by the weatherperson to confuse you; it means that the wind pattern is such that the front moves in one direction on one part of the chart, and in the opposite direction elsewhere. And almost by definition, that which is a warm front moving east, will be a cold front if it moves west.

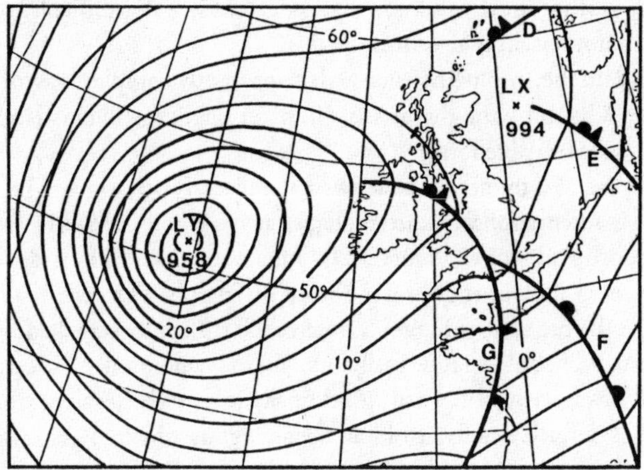

A typical North Atlantic weather-chart, showing a depression to the west, an occluded front over the north of Ireland, and a warm sector over France. (*Weather*, 42, 1987)

Odd Numbers

Every hour, precisely on the hour, weather observers at many hundreds of meteorological stations around the world takes careful note of the prevailing weather conditions. They use instruments to measure some of the more important weather elements, and then compile all the available information into a special coded message. The completed report is fed into a dedicated world-wide telecommunications network to form the raw material for tomorrow's forecast. It might look something like this:

221200 03953 41360 53110 10067 20023 49968 52017 69921 78011 83710 85820 =

Now, looking at this jumble of numbers, one might be reminded of the young Pip in Dickens's *Great Expectations*, who nightly 'fell among those artful thieves, the nine figures, who seemed every evening to do something new to disguise themselves and baffle recognition'. But these numbers allow forecasters a great distance away to build up a precise picture of the weather they describe.

The first six-figure group is the time and date – in this case 1200 on the twenty-second. In the second set of numbers, '03' tells us that the weather station is in Ireland, and '953' is the identifying number assigned to Valentia Observatory in Co. Kerry. All the other figures give precise information about different aspects of the weather – the visibility, the type and amount of cloud and the height of the various layers above the ground, the temperature and humidity, the speed and direction of the wind – and so on. If you know the code, and it is not difficult to learn, all this information is at your fingertips.

Why do meteorologists choose this apparently complex way of doing things? Why not write the message in plain English, and circulate it in a much more readable form?

The use of a numerical code has a number of advantages. The first is that it is international: a meteorologist anywhere in the world can read the report and know immediately where it came from and what it means, even if he or she has not a word of English. The code is also very concise: the message in figures is much shorter than its verbal equivalent, and can be handled more easily and more economically by computer than if it was written in words. And finally it is precise: each of the code-figures is defined exactly, avoiding the ambiguity which might arise from using common words with differing shades of meaning.

The World in a Circle

William Blake's ambition to 'see a world in a grain of sand' requires the kind of apocalyptic imagination which is possessed by very few. But any forecaster can deduce the weather over the whole of Europe from a jumble of figures and dots. This feat implies no transcendental vision on his or her part; it is a happy consequence of the concise and orderly presentation of information.

A weather-chart ready for analysis by the forecaster is covered with freshly plotted numbers and symbols, carefully arranged in clusters of about half an inch square. Each cluster summarizes the weather at a particular spot on the map. On a large-scale chart of Europe there may be as many as a dozen such groups of figures plotted over the outline of Ireland; off the coast there may be similar groups depicting observations from ships at sea, oil rigs, weather buoys or lighthouses.

The station of origin of each report is pinpointed by a little circle. The numbers and symbols are arranged around the 'station circle' in a carefully chosen pattern which corresponds to the points on an eight-point compass; their meaning can be deduced from their position and from their colour, which may be red or black.

The barometric pressure is plotted in the 'north-east' corner of the cluster. Just below it, immediately to the right of the station circle in the 'east' position, is the pressure tendency – the amount by which the pressure has risen or fallen during the previous three hours. Further down, in the 'south-east' corner and plotted in red, is a symbol which signifies the 'past weather' – what the weather has been like for the six hours prior to the time of the report. A single dot indicates rain, a comma is used for drizzle, a triangle for showers, a star for snow, and other signs for any kind of weather you might think of.

And so it goes, right around the station circle. Temperature, humidity and wind, together with details of the height, amount and types of clouds are all included in this compact ensemble of meteorological hieroglyphs. When the chart is plotted, it is possible with a little practice to assimilate at a glance the weather pattern over a whole continent.

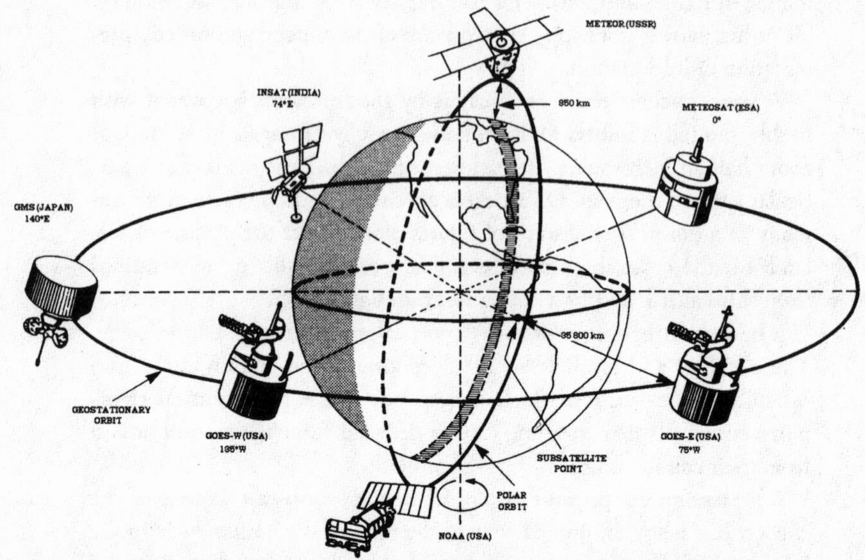

Diagram of the World Meteorological Polar-orbiting and Geostationary satellites. (WMO 25th Anniversary booklet)

The Endless Isobar

'Some words', wrote W.S. Gilbert of Gilbert and Sullivan fame, 'are full of hidden meaning – like Basingstoke.' The word 'isobar' might be described as such. There is more to it than meets the eye.

Looked at in one way, there is no such thing as an isobar. You cannot see it or feel it, or come across it in any tangible form in real life. It exists only in the imagination, a theoretical concept introduced as a mental crutch, to help assemble thousands of values of atmospheric pressure into some form of recognizable pattern. In this context it is different from fronts, which are equally prominent on the weather-map; a front has a real existence as a boundary between two masses of air, and its physical presence is often marked by clouds and heavy rain.

An isobar is a line on a weather-chart drawn through points of equal atmospheric pressure. Etymologically, it comes from a combination of two Greek words: *isos*, meaning 'equal', and *baros* meaning 'weight'. The name makes sense if you remember that the atmospheric pressure at any point on the earth's surface is the weight of the column of air vertically above that point, right up to the very top of the atmosphere.

As part of the regular sequence of detailed weather observations, the forecaster receives hourly values of atmospheric pressure – or *barometric* pressure, as it is often called – from many hundreds of weather stations around the world. When these pressure values are plotted on a chart, he or she can draw lines on the chart joining points of equal pressure. The technique is similar to that used by the cartographer, who draws height contours on a conventional map joining points which are the same distance above sea level. Indeed the finished chart is closely analogous to a map of the physical features of a given area – except that instead of mountains and valleys, the isobars identify areas of high and low pressure – anticyclones and depressions.

On maps covering a large area, isobars are normally drawn at four millibar intervals. Since each line represents a specific value of pressure, isobars never cross; if they did it would imply two different pressure values at the same point, which is clearly an impossibility. And an isobar never ends, except at the very edge of a chart; indeed if isobars were drawn on the globe, each and every one of them would be a continuous closed curve.

Getting the Isobars Straight

An isobar, well drawn, is a thing of great beauty to a meteorologist. He sees it as his personal work of art, a thin black line, sweeping its way across the chart in graceful curves and slowly, gently changing course to follow the local flow of wind. A single isobar may appear on the map near Newfoundland, cross the frozen Arctic waste and then veer southwards towards Berlin by way of Glendalough and Isles of Scilly; thence, perhaps, it may make a sudden sally into Greece before passing into *terra incognita* beyond the chart's edge.

The whole ensemble is made possible by noting the pressure values reported from the hundreds of weather stations over the thousands of square miles represented on the chart, but the concept is less simple than it might seem to be at first. Pressure varies considerably in the vertical; it decreases by about one hectopascal for every 30 ft in height, a much greater rate of change than often appears over considerable distances in the horizontal. But weather stations differ considerably in their altitude; if we were to examine a chart on which were plotted the atmospheric pressure as read directly from the barometer at each station, no recognizable pattern would emerge, because of random differences in the heights of stations.

To avoid this difficulty, the pressure values from all stations must refer to a common level – the chosen one being Mean Sea Level. For the mariner, this makes matters easy; he simply reads his barometer where he is and that is an end to it. The land-based observer, however, must be more circumspect. The pressure read at each station must be 'reduced' to a value appropriate to sea level immediately beneath it assuming that air, rather than rock or soil, filled the intervening space.

The correction is accomplished by adding to the pressure reading obtained from the barometer an amount equal to the weight of an imaginary column of air extending from barometer down to sea level. This is not difficult to calculate, and depends mainly on the air temperature at the time of the reading. In practice, the required corrections are worked out for each station for a complete range of possible temperatures, and all the observer has to do is to read the appropriate figure from a card.

Perplexing Winds

One of the more surprising features of the weather-map is the fact that the wind blows *along* the isobars. At a first guess, one would expect air to flow from an area where pressure is high, directly to a region where pressure is relatively low – which would result, in the wind blowing *across* the lines of equal pressure. And this indeed would be the case if the earth were 'tideless and inert', if it were smooth, flat and stationary.

But our world revolves. We are, as Wordsworth put it, 'rolled round in earth's diurnal course', on a planet which is nearly spherical and rotates on its axis once every twenty-four hours. It is this complicating factor which makes the wind behave in its rather perplexing way.

To understand the effect of the earth's rotation, one must recognize that all motion is *relative*; its appearance depends on the point from which it is observed. Imagine, for example, a rocket launched southwards from Donegal towards Cork. According to Newton's First Law of Motion, it will continue on a straight line unless some other force changes its direction. Since – insofar as it affects this argument – no such force exists, the rocket, looked at from outer space, would be seen to follow a straight course.

But as our rocket moves southwards over Ireland, the earth is turning underneath it from west to east. Instead of striking the ground somewhere in Co. Cork, it would land at a point significantly to the west, perhaps on an unpleasantly surprised Dingle! To *an observer on the ground*, it would appear that the trajectory of the missile curved towards the west, a phenomenon well known to meteorologists as the *Coriolis Effect*.

Air, despite its insubstantial nature, is continually subject to the Coriolis Effect. The rocket example above is merely a very simple illustration of its consequences; the full mechanism is a great deal more complex, and results in the general rule that air moving over the surface of the earth in the northern hemisphere is deflected to the *right*. It may start to blow *across* the isobars from high to low pressure, but changes direction until it ends up blowing *along* the isobars with low pressure on the left-hand side. This strange behaviour of the wind was first explained in 1835 by a French engineer and mathematician called Gaspard de Coriolis, from whom – if one may risk being obvious – the phenomenon takes its name.

A Useful Idea

Over a thousand years ago, in the middle of the ninth century, Pope Nicholas I made a significant contribution to meteorology. It was he who decreed that the figure of a cock should surmount the topmost pinnacle of every abbey, cathedral and parish church in Christendom.

The emblem was intended as a reminder of St Peter's weakness. Having denied the Lord three times, Peter heard the crowing of a cock, and was reminded of his master's forecast, whereupon, we are told, the repentant apostle 'went out and wept bitterly'. The papal ordinance was harnessed to another purpose by mediaeval architects who used the 'weathercock' to show the wind's direction.

But weathercocks have been superseded; today forecasters inspect the isobars on weather-charts, and use them to deduce the strength and direction of the expected wind.

The wind, as we have seen, tends to blow along the isobars. Indeed if the atmosphere were to reach perfect equilibrium, the wind would blow exactly along these lines of equal pressure, a hypothetical situation that meteorologists imagine by the concept of the *geostrophic wind*. It is only an ideal, because in practice the slowing down of the wind by rough terrain, and the fact that pressure systems are constantly developing and decaying, cause the atmospheric system always to be slightly out of equilibrium. The geostrophic wind is useful as a concept though, since it can be easily measured from a weather-chart, and serves as a good approximation to the real thing.

The speed of the geostrophic wind is inversely proportional to the distance apart of the isobars; the closer they are together, the stronger the wind will be. To find the wind at a particular spot on the weather-chart, the forecaster first measures the perpendicular distance between two adjacent isobars; he or she then notes the latitude, and reads the appropriate geostrophic wind speed from a special scale. From experience, he or she then adjusts this theoretical value to arrive at an estimate of the actual wind at the particular place in question.

Direction Signs

It is the forecaster's task, as he or she analyses the weather-chart, to seek the *concordia discors*, the harmony hidden beneath the apparent discord. A myriad of scraps of information – many of them in seeming contradiction – must be pieced together, each with its correct emphasis, and woven into a thread of logic which will ultimately reveal the way in which the pattern of fronts and isobars is most likely to evolve. Take, for example, the way in which the forecaster estimates the future position of a typical North Atlantic depression.

Rules of thumb can give initial guidance. The forecaster first looks at the looks at the history of a depression, and by examining a number of charts, drawn at, say, six-hourly intervals, the progress of each depression can be followed. There is no guarantee that each one will continue to move in the same direction, or at the same speed, but 'What's past is prologue' and history at least sets the scene.

Next, the forecaster will bear in mind that depressions frequently move in the direction of the *warm sector* isobars. The warm sector is the triangular area between the warm front (with the round barbs) and the cold front (with the little triangles). The isobars in this region tend to be roughly parallel to each other, and their orientation is a good 'first guess' as to the direction in which the parent depression is likely to travel.

Further clues are to be had from the pressure 'tendencies' plotted on the chart, the amount by which the pressure at each weather station has changed in the previous three hours. Pressure falls most rapidly at stations in the direct path of a low, so by noting where the falls are greatest, the forecaster can estimate the likely direction of movement of a depression.

The forecaster is also aware that depressions tend to be 'steered' by the winds in the upper levels of the atmosphere. The winds at about 20,000 ft are often a good guide, and their direction and speed in the vicinity of the surface depression often betray the future plans of the low underneath.

All these clues are important but they may provide conflicting advice. The measure of a good forecaster is the degree to which he or she can sift the available evidence, and apply skill, experience, and scientific knowledge, giving each clue the importance it deserves, to arrive ultimately at the right answer.

Iso-Eccentricity

Edmond Halley was a man of many talents. A close friend of Isaac Newton, and Astronomer Royal at London's Greenwich Observatory, Halley is best remembered for his 1705 prediction that the comet which now bears his name would make a reappearance in the year 1758. But he has another claim to fame: he is credited with inventing the *isopleth*.

An isopleth is a line drawn upon a chart which joins points having assigned to them the same numerical value of any given parameter. There are many kinds of isopleth. Most have names which begin with the Greek prefix *iso-*, meaning 'equal'; the *isobar*, as we have seen, is a line joining points of equal pressure. But there are many, many more.

Halley's original isopleths had nothing to do with meteorology. The idea came to him during a long voyage in the Atlantic from 1699 to 1701, during which he carried out an extensive series of observations of the earth's magnetic field. These enabled him to draw isogonic lines, which joined points of equal magnetic declination. Scientists in general, and meteorologists in particular, have been drawing isopleths ever since, plying their art 'in liquid lines mellifluously bland'.

Meteorologists love isopleths! They draw them quite compulsively at anytime and anywhere they can, and the opportunities to indulge themselves in this way are legion in their walk of life. They draw *isotherms*, for instance, to join points of equal temperature on their charts. If they are in a statistical mood, they draw *isocrymes*, which join points of equal mean winter temperature, and *isotheres* to show the pattern of mean summer temperature. The lines they etch on rainfall maps are called *isohyets*, and those which show the distribution of recorded sunshine around the country, *isohels*. A complementary pattern emerges as they examine the *isonephs*, the lines of equal cloud cover.

In the case of the wind, the lines which matter are those which join the points where the speed of the wind is uniform, the *isotachs*. *Isochrones*, on the other hand, are lines of equal time, often used to show successive hourly positions of a front on a weather-map. And after a thundery spell, the avid weather person may sometimes be found analysing the *isoceraunic* pattern: that which shows the percentage of days within a given period on which a peal of thunder has been heard.

The Height of Fashion

'For an idea ever to become fashionable is ominous, since it must ever afterwards be considered old fashioned.' This at least was the view of the American writer George Santayana. And thus it was with the use of kites in meteorology. They were all the rage around the turn of the century, and for a few years around that time they provided the best available method of gathering information about the thermal structure of the atmosphere high above the ground. Nowadays they are a mere curiosity.

During most of the nineteenth century, the only way in which scientists could discover the nature of the upper atmosphere was by the somewhat dangerous expedient of ascending in a manned balloon. But manned balloons were limited in the heights they could reach; for one thing, the aeronaut suffered from lack of oxygen if he went too high. Then, in the early 1890s, the *meteorograph* was invented – a self-recording device which provided a continuous trace of temperature, pressure and humidity. One might say, to coin a phrase, that it opened new horizons in the vertical.

With this innovation, balloon ascents were quickly abandoned, and for the next ten years or so, everyone was using kites. The kites were of the 'box' type, and the cable consisted of a single strand of steel piano wire. Kites would be strung out at intervals of a mile or so along its length, and by this means it became possible to raise the meteorograph to heights of three or four miles; on one occasion a height of over 30,000 ft was reached. A powerful winch, sometimes driven by a small steam engine, was often used to control the whole assembly.

By the middle of the 1890s kite ascents had become routine at many observatories around the world. Many were scheduled for 'International Days' to allow simultaneous – or *synoptic* – observations to be made, and making possible the detailed analysis of the vertical structure of the atmosphere over a large area.

But by the early years of the twentieth century it became evident that there was a limit to the height which kites could reach. An alternative method gained in popularity; the new technique of sending the meteorograph aloft on an unmanned hydrogen-filled balloon. Although this method had one serious disadvantage – its success depended on someone finding and returning the instrument at the end of the flight – it nonetheless became the standard method of carrying out upper-air observations until the development of the *radiosonde* during the 1920s.

Eyes Up

A meteorologist must be high-minded. This is not to say that he must constantly appropriate the high moral ground, or that – in the words of W.S. Gilbert – he must be always 'anxious for to shine in the high aesthetic line as a man of culture rare'. It means, rather, that he must be aware of what is going on above his head.

A factor of major importance when trying to assess the future behaviour of the atmosphere is the pattern of winds at upper levels. Information about this pattern is obtained by using a hydrogen-filled balloon which, as it ascends, is carried along horizontally by the wind. A series of spot readings of the balloon's position gives sufficient information to calculate the average speed and direction of the wind in each layer of the atmosphere.

In the early days of meteorology a 'pilot' balloon was used for this purpose. It was a small rubber balloon about three feet in diameter, filled with sufficient hydrogen to cause it to rise at a more or less constant rate of about 500 feet per minute. During the ascent, its progress was followed with a *theodolite* – a kind of surveyor's telescope which was kept pointing at the target, and from which angles of azimuth and elevation could then be accurately read. A special slide-rule translated these angles into a value for the average wind between each two successive readings. The time which had elapsed since the launching of the balloon gave the approximate height to which the values applied.

With favourable conditions, wind measurements at levels up to 20,000 feet or more could be obtained using a pilot balloon. But the method had a number of serious disadvantages. Its use was severely restricted by low visibility or by the presence of cloud while very strong winds would carry the balloon quickly out of sight before it had time to rise more than a few thousand feet.

During the 1940s, the advent of radar provided a much-improved method. A metal reflector attached to the ascending balloon allowed the radar operator to keep the antenna pointed in its direction; once again azimuth and elevation could be used to calculate the upper winds. Moreover, the radar also gave an accurate assessment of the balloon's height above the ground, so there was no guesswork as to the height to which the calculated winds applied. Nowadays, the latest satellite-based navigation systems continuously pin-point the exact position of the balloon, allowing upper level winds to be obtained with very great precision.

High Winds

Although there are local eddies in the flow, our atmosphere – broadly speaking – drifts eastwards around the earth, resulting in winds which are more or less westerly in direction. But at any particular time it is usually possible to identify clearly a most interesting phenomenon – a thin core of very strong winds high above the earth, winding its way with snake-like undulations around the globe. This is the *jet stream.*

The jet stream may be anywhere between 25 and 40 thousand feet above the ground. The narrow tube of strong winds is normally 100 miles or so in diameter, and typically blows at a speed of 100 to 150 miles per hour – although winds as strong as 300 m.p.h. have been experienced near the core on a number of occasions. This slender river of swiftly moving air is in sharp contrast to the much quieter atmosphere surrounding it; wind speed drops off very sharply at its edge, and also above it and below.

The jet is orientated in a roughly west-to-east direction, although it often meanders in a wavy U-shaped pattern. It is caused by sharp contrasts of temperature in the horizontal, a sudden lurch in the otherwise gradual decrease in temperature with latitude. The temperature drop – and hence the speed of the jet – is greater in winter than in summer; in January, for instance, the temperature difference between the equatorial and the Arctic regions can reach 70°C or more, while in July it is only about half this amount.

The jet stream is of great importance to the aviation industry, since high-flying aeroplanes going from west to east can save a great deal of time and fuel by taking advantage of its strong winds. An aircraft with a cruising airspeed of 550 m.p.h. might take seven hours to fly from New York to Paris in normal conditions; if it flies for much of the time in a jet stream blowing at 150 m.p.h., more than an hour's flying time will be knocked off the journey. But naturally, on a flight from east to west, a pilot would seek to give the jet stream a wide berth.

There is sometimes a price to be paid for taking advantage of these strong tailwinds. Jet streams are often turbulent, because of sharp variations in wind speed 'near to the river's trembling edge', and the result can be a violent buffeting of the aircraft as severe as would be expected in the worst of thunderstorms. Since this usually occurs with cloud-free skies, it is called *clear air turbulence* – or CAT for short.

Points of View

The mathematics of putting an artificial satellite into orbit around the earth were known for a long time before that feat was finally accomplished. The great difficulty was the rocket power with which to achieve it. When the threshold of space had been crossed with the launch of Sputnik 1 on 4 October 1957, scientists began to be choosy about where to put their satellites. Many kinds of orbit are possible, and the right one in a particular case depends upon the purpose for which the satellite is intended. Two kinds of orbit are popular with meteorologists.

One is a *geostationary* orbit. Someone, sometime, had the bright idea that if the speed of a satellite in orbit over the equator were to be synchronized exactly with the rate of rotation of the planet, the satellite would appear to be fixed in space. It would look down constantly at the same segment of the globe, and successive pictures taken by it at, say, half hourly intervals, could be combined to form a 'movie' of the evolving weather situation. To achieve this happy state of affairs, it is necessary that the satellite be in orbit some 23,000 miles above the equator.

But this type of orbit has disadvantages. The European geostationary satellite METEOSAT, for example, is positioned over the Gulf of Guinea, which lies just south of Nigeria. It provides an excellent view of Africa and of the southern states of Europe, but northern countries are seen at a very oblique angle because of the curvature of the earth. Moreover, owing to its great distance from the ground, temperature values from METEOSAT and other data used for numerical weather prediction, are less accurate than scientists would like.

Polar-orbiting satellites can cope with these difficulties. As their name implies, they travel around the earth from pole to pole, following, as it were, the lines of longitude. They have a much lower orbit – about 500 miles above the earth – and are therefore in a position to obtain more accurate data than the geostationaries. And far from neglecting the countries in northern latitudes, a polar-orbiting satellite gives them special attention, since the lines of longitude which it follows converge towards the poles.

Unfortunately, however, each picture from a polar-orbiting satellite is taken from a different vantage point. As a consequence, the images of a sequence cannot be compared with one another as easily as those from a satellite in a geostationary orbit.

Photograph taken by a METEOSAT Geostationary Satellite over the intersection of the Greenwich Meridian and the Equator. (Courtesy of EUMETSAT)

Ahead of His Time

L.F. Richardson is remembered by meteorologists as the father of numerical weather prediction: weather forecasting by computer. As far back as 1922, in a famous book called *Weather Prediction by Numerical Process*, Richardson outlined a method by which the future pressure pattern could be calculated if the present state of the atmosphere were accurately known. He devised a number of equations which encapsulated many of the known principles of physics – like Boyle's Law and Newton's Laws of Motion – and his forecasting technique consisted of applying these equations repeatedly to advance the forecast in short time-steps.

The theory seemed plausible but it was a method without a means. An intimidating amount of calculation was involved, and in those pre-computer days there seemed to be no practicable way in which the technique could be implemented.

With tongue in cheek, Richardson made a suggestion. He reckoned he needed a 'weather factory' with 64,000 workers, whom – somewhat prophetically – he called 'computers'. 'Imagine a large hall', he wrote, 'like a theatre, except that the circles and galleries go right through the space usually occupied by the stage. The walls of this chamber are painted to form a map of the globe, and a myriad of computers are at work upon the weather at the part of the map at which each sits.' The whole operation is under the baton of a supervisor who sits in a large central pulpit. His function is to 'maintain a uniform speed of progress in all parts of the globe, being rather like a conductor of an orchestra in which the instruments are slide-rules and calculating machines.'

In a more serious frame of mind, Richardson tested his technique by painstakingly working through the calculations to produce a six-hour forecast. It took him several months, and the result was disappointing; it predicted pressure changes of up to 150 millibars in the six-hour period. But Richardson has been vindicated. Some time ago Dr Peter Lynch of the Irish Meteorological Service 're-ran' Richardson's forecast on a computer, using the original data and the equations devised by Richardson himself. The important difference was that Lynch used a modern technique called 'initialization' to smooth out tiny inconsistencies in the original observations. With this improvement, Richardson's method works, and gives a reasonable forecast – no mean achievement for a man who was forty years ahead of his time.

above: Artist's impression of Richardson's 'Weather Factory'. (A. Lannerbach, Dagens Nyheter, Stockholm)

Lewis Fry Richardson (1881–1953), father of modern Numerical Weather Prediction. (From *Prophet or Professor?* by O.M. Ashford)

Doing It by Numbers

With the arrival of satellite pictures, many people thought that the problems of forecasting the weather were over: now that the forecaster can see exactly what is happening, surely there is no excuse for ever being wrong again?

Life, unfortunately, is rather more complex than that, or as Oscar Wilde put it: 'The truth is rarely pure and never simple.' The satellite pictures provide information about the weather pattern as it appears now; they do not say what the picture will look like in six hours time, or two or three days ahead. Other methods must be used to solve this problem, albeit that this information from the satellite is useful as a starting point. In this respect, computers have had a quite dramatic impact on meteorology.

Forecasting by computer is based on a mathematical model of the atmosphere, a description of the behaviour of the atmosphere in terms of mathematical equations. Scientists know that air – as a gas – obeys certain physical laws; it expands, for example, at a certain rate when heated by a given amount. It is also possible to describe mathematically the ways in which energy changes from one form to another, say from heat into motion, or vice versa.

Armed with the equations which specify these processes, and given the values of pressure and temperature at a certain spot in the atmosphere, expected values of these parameters at some future time can be calculated. If the same operation is carried out by computer for hundreds of points on the weather-chart, a new weather-map for some future time can be constructed. The computer moves forward step by step, hour by hour, until it arrives at a forecast four or five days ahead of the original observations.

The models used in numerical weather prediction are very complex, and are continually being refined and improved to reflect the behaviour of the real atmosphere with greater accuracy. Although the human weatherperson is still better than the computer as regards very short-range forecasting, in recent years he is no match for the machine when it comes to predictions for several days ahead. His expertise now lies in the skillful interpretation of the computer products; he must translate those mysterious lines on the chart produced by the computer into a description of the weather which will be meaningful to the man on the Dalkey omnibus.

Checking Up on the Weatherman

'There is nothing either good or bad', declared Hamlet, 'but thinking makes it so.' As it happened, he was defending a hastily formed view that 'Denmark's a prison', but he might as aptly have been speaking of weather forecasts. It is not always easy to tell a good one from a bad one!

To the man in the street a forecast which turns out to be correct is always a good one. But accuracy is no guarantee of skill on the part of the forecaster; skill is only in evidence if the prediction is correct 'against the odds', as it were. Let me illustrate the point with what I hope is a simple example.

During the twenty-five years from 1960 to 1984 ground frost occurred at Kilkenny in August on a total of fifteen occasions; this makes 15 days out of 775, or about 2 per cent. Anyone aware of these figures could predict 'no frost' for Kilkenny each day during August in any particular year, and be confident of being right about 98 per cent of the time. Although the percentage accuracy is high, the forecast is not impressive; to earn his or her keep, the weatherperson must be able to do better than merely use climatology, and should be able, for example, to predict those rare days in August when frost actually *will* occur.

To reassure themselves about their usefulness, meteorologists devise *objective* ways of measuring the skill of their weather forecasts, like this, for example: imagine that in a certain place, the climatological records show that on average it rains on ten days in the month of September; a 'forecast' based purely on climatological expectations can be devised by randomly selecting ten dates from the month, and 'forecasting' rain on those days. In due course, the weather forecaster using experience and all the advanced facilities that a modern weather office has to offer, produces his or her own prediction for each day of September.

At the end of the month the results of both sets of forecasts are compared with the actual observations of days with rain. With this information, one simple 'skill score' is the ratio of the correct forecasts produced by the two procedures. Let me explain: if the forecaster, for instance, were correct on twenty-five days out of the thirty, and the climatological prediction were correct on fifteen days, the skill could be represented as 25/15, or 1.66. The greater the value of this ratio, the more skillfulthe scientific forecasts. A useful application of this technique is that over a long period, meteorologists have a useful measure of whether their forecasts are improving, or not.

A Recipe for Air

Pure fresh air is a great tonic. But to speak of *pure* air is almost a misnomer, since air is not a specific gas in its own right. As Shelley so aptly put it, no doubt without really knowing he was so much to the point, 'The winds of heaven mix for ever;/Nothing in this world is single.'

Air is a mixture of many gases. The vast bulk of them remain more or less fixed in proportion to each other, proportions which do not change significantly up to a height of fifty miles or so above the earth. A tiny part of the atmosphere comprises constituents which vary in quantity from place to place, and from time to time.

Nitrogen predominates in the atmosphere, and accounts for about 78 per cent of dry air. It is generated mainly by the decay of agricultural debris and animal matter, and is exuded into the atmosphere by eruptions from volcanoes; certain rocks release nitrogen, as does the burning of some fuels. On the other side of the ledger, nitrogen is removed from the atmosphere by the biological processes of plants and sea life; to a lesser extent, lightning and high temperature combustion convert nitrogen gas to nitrogen compounds, that are washed from the atmosphere by rain and snow. Production and destruction of nitrogen are in balance, so the quantity in the atmosphere remains constant.

Oxygen, the second most abundant gas, is crucial to human and animal life, and accounts for slightly less than 21 per cent of air. It is produced by vegetation, which takes up carbon dioxide through photosynthesis, and releases oxygen. It is removed from the atmosphere by man and animals, whose respiratory systems are the reverse of those of the plant community; oxygen also dissolves in lakes, rivers and oceans, where it serves to maintain marine organisms, and is consumed in the process of decay.

Not many years ago it was feared that the widespread reduction in plant life, and other man-made environmental changes, might threaten the supply of atmospheric oxygen. Nowadays, however, scientists are happy – whatever our other environmental problems – that the supply of oxygen is remaining more or less constant.

A little mental arithmetic reveals that these two gases account for about 99 per cent by volume of the air we breath. Other gases are present in only tiny quantities, although as we know from current concerns about climate change and the ozone problem, their importance in the scheme of things well belies their tenuous existence.

The Right Site

Much of the information contained in a standard weather report is obtained from meteorological instruments. Nothing, you might think, could be more simple: to find the temperature at a particular spot, for instance, you put a thermometer there and read what it says. But the exercise requires careful planning. It is important that meteorological instruments be positioned correctly if the information they will provide is to be of any use.

Temperature, to pursue the example, changes very rapidly with height in the first few feet above the ground. If temperature values for different places are to be comparable, they must be taken at a *standard* height – which meteorologists have agreed to be four feet above ground level. Also, the thermometers must be positioned in the shade; it is the temperature of the *air* which is required, not the artificially high temperature of the thermometer itself when heated directly by the sun.

Meteorological thermometers are housed in a louvered box called a Stevenson Screen. The screen is painted white to reflect the sun's rays, and the louvres allow the air to circulate freely through it. It is constructed in such a way that the thermometers inside are positioned at the correct height above the ground.

Barometers, on the other hand, are normally kept indoors, but they too must be placed in a position where they are not affected by direct sunlight. The heat of the sun would cause the mercury to expand, resulting in spurious pressure values. The height of the instrument is not critical; pressure readings are adjusted for elevation, to give a figure for what the pressure *would* be if the barometer were positioned at Mean Sea Level.

The requirements for measuring the duration of bright sunshine are rather obvious. Any shadows which might fall on the sunshine recorder from nearby buildings or mountains will affect the record, so the instrument must be sited in such a way as to avoid such obstructions, or at least to minimize their effects.

When measuring the wind, height is again an important factor. Meteorologists have agreed that the standard height at which wind should be measured is thirty-three feet above the ground. The anemometer must be sited on flat open ground, to ensure that the flow of wind in the vicinity is as smooth as possible. Buildings or trees of any size, even quite a distance away, cause turbulence in the airflow and make wind readings unrepresentative.

Applying Pressure

The story of the barometer, the instrument used by meteorologists for measuring atmospheric pressure, could be said to have begun with the birth of Evangelista Torricelli on 15 October 1608. He was born in Italy in a little town called Faenze, but early in his career he moved to Florence where he became a close associate of the great Galileo Galilei. In due course he was appointed Philosopher and Mathematician to Grand Duke Ferdinand II of Tuscany, a post he held until his death in 1647.

Torricelli was recognized at a personal level as a man of great charm and modesty. He was also one of the foremost mathematicians of his day, and his place in history was assured by the so-called Torricellian Experiment performed in 1643, essentially the invention of the mercury barometer.

Torricelli's barometer consisted of a glass bulb fitted with a neck 'two cubits' long – about forty inches. The tube was filled with mercury and inverted into a dish containing more mercury, whereupon it was found that the mercury in the glass bulb and tube fell to a certain level – about thirty inches above the surface of the liquid in the dish – but no further. Close observation showed that the height of the column of mercury varied with changes in the weather.

Other scientists of the day were quick to build on the foundation laid by Torricelli. Five years later, in 1648, the Frenchman Blaise Pascal induced his more agile brother-in-law, Florin Perrier, to carry a mercury barometer up the 5000-foot Puy-de-Dôme, the highest peak of the Massif Central. The decrease in atmospheric pressure with height was demonstrated by showing that the mercury column was shorter at the top of the mountain than it was at the base.

Then, during the 1650s, Otto von Guericke put the barometer to practical use. He made a 'water barometer', which because of the relative lightness of the fluid required a much longer tube – over thirty feet in length. He attached this tube to the side of his house in Magdeburg, so that the passing public could observe the figure of a little man on a cork floating at the top of the column; it floated high when the weather was fine, but dropped with the level of the water at the approach of stormy weather. By his novel and popular application of the barometric principle, von Guericke might well be considered to have established the first ever scientific Weather Centre.

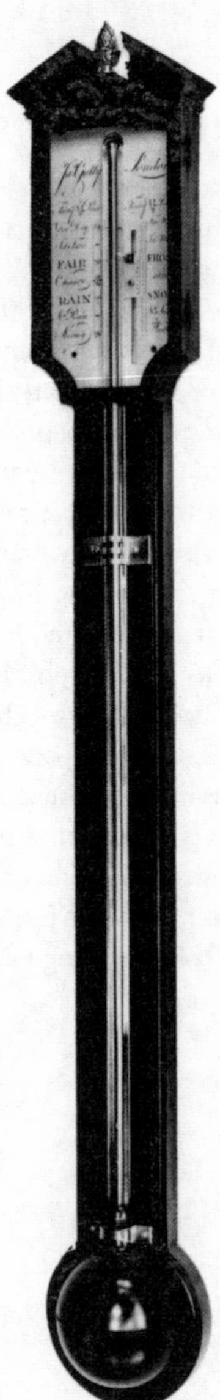

Jas Gatty, London
38·5″ × 5·2″
City of Gloucester Museum: from the collection of S. Marling

Highly decorated early nineteenth-century Mercury Barometer. (City of Gloucester Museum)

Setting the Trends

The ordinary domestic barometer has a dial which provides dire warnings of 'rain' and 'storms' on its left-hand side, and moves clockwise to cheerful intimations of 'bright' and 'sunny' on the extreme right. But these suggestions are based on the assumption that the needle indicates the *sea-level* pressure. If the instrument has not been correctly set, its premonitions will be worse than useless.

A new barometer, in all probability, will be factory set, and if your hallway happens to be at a significantly higher or lower level than that factory, wherever it may have been, the instrument will read incorrectly. You will recall that pressure decreases rapidly with height – at a rate of about 1 millibar for every 30 ft. This means that a difference in altitude of a mere 300 ft between the level where the barometer was originally calibrated, and its eventual position on the wall *chez vous*, will result in a discrepancy of 10 millibars. The instrument may tell you what the atmospheric pressure is in your hallway, but its relevance as regards the weather patterns described on the dial will be severely distorted.

The solution is easy. A knob at the back of the instrument allows you to change the current position of the needle to indicate the local sea-level pressure, and your nearest Meteorological Station will usually be glad to oblige with the correct setting for your locality.

The Late Millibar

The millibar died officially on 1 January 1986. From the beginning of organized meteorology it had been the internationally recognized unit for the measurement of atmospheric pressure. Then, with a single stroke of the pen, it was gone! Not quite gone, of course; it is still used informally for many purposes, and indeed for the most part throughout this book. But in all official meteorological publications, the *hectopascal* – sometimes abbreviated to 'hPa.' – is now the approved term.

Atmospheric pressure and barometric pressure are just different names for the same thing. We live, as it were, in a sea of air, in much the same way as underwater creatures live in a sea of water, and any area of the earth can be thought of as supporting the weight of the overlying atmosphere. The pressure at any point on the earth's surface is the weight of the entire column of air, right up to the top of the atmosphere, standing on the unit area at that point. The pressure varies from place to place, because of variations in temperature – and hence the density and weight – of the air above it. The average value of atmospheric pressure is about 14.7 lbs per square inch.

It was the French novelist Alphonse Karr who first remarked that 'Plus ça change, plus c'est la même chose' (The more things change the more they remain the same). And so it was with the change in pressure units – although at first glance the distinction between the millibar and the hectopascal seems formidable.

A millibar is one thousandth of a *bar*, which is the basic unit of pressure for meteorological purposes in the old centimetre-gram-second, or *cgs* system of scientific units. But nowadays in scientific circles, the *Système International* – the SI system of units – is almost *de rigueur*, and meteorologists had to follow suit. Pressure in this system is measured in *pascals*, and for meteorological purposes the most conveniently sized multiple is 100 pascals, the *hectopascal*.

Happily, and not entirely by coincidence, a hectopascal is numerically equal to a millibar. Like atmospheric and barometric pressure, they are just different names for the same thing, so a depression of 995 millibars conveniently turns out to be also a depression of 995 hectopascals.

Points for Good Weather

'Comparisouns doon offte gret greuaunce', remarked the mediaeval writer John Lydgate, or as it was more succinctly put by Dogberry in *Much Ado About Nothing*, 'Comparisons are odorous.' In the pursuit of science though, comparisons are often useful. Over the years meteorologists have spent a great deal of time trying to devise ways of comparing one summer with another.

All of us remember good and bad summers in our own subjective ways. But our views may be coloured by what we were doing that year, or by how much time we had to enjoy such good weather as there may have been. On the other hand, *objective* measures of good weather are hard to formulate; a warm summer that is wet and dull is no less mediocre than a bright one with chilly winds. A yardstick of good weather must account for all the factors that might make it pleasant or otherwise.

Some thirty years ago an English meteorologist called R.M. Poulter did precisely that. He came up with a *Summer Index*, a system of points which allotted a score to each summer, thereby making it possible to compare one with another.

The Poulter Index works on the assumption that the character of a summer depends mainly on three specific parameters: temperature, rainfall, and sunshine. Other features – such as cloudiness, windiness, or the frequency or absence of thunderstorms – are related to these elements, and in any case they have a much smaller impact on our collective memory. The score assigned to each summer is given by the formula:

$$I = 10T + S/6 - R/5$$

Put simply, this score is based on *ten* times the average temperature for the three summer months (T) added to *one sixth* of the total number of hours of sunshine (S), from the sum of which is subtracted *one fifth* of the June, July and August rainfall (R), measured in millimetres. The answer is the Poulter Index (I) for the summer in question.

The fractions and multiples included in the formula are specially calculated to give approximately equal weight to each of the three components. The result of these calculations is a number which is an indication of how good, or bad, that particular summer was. The summer of 1976, for instance, had a Poulter Index in the midlands of Ireland of 405, while that of 1985 scored a miserable 249.

III *Coming to Blows*

Down in the Doldrums

When the great voyages of discovery began in the middle of the fifteenth century, the open ocean was regarded by mariners in much the same way as space must appear to present-day astronauts, except that the sailor knew less about where he was going and had fewer hopes of returning home.

He was afraid to sail down the Atlantic coast of Africa beyond Morocco for there he would enter the 'green sea of darkness', a perilous swamp full of monsters, reliably documented by Arab geographers; if he headed north, he would find himself in a frozen waste, an 'undiscovered country from whose bourn no traveller returns'; and if he sailed westwards across the Atlantic he might drift close to the equator where men turned permanently black and no life could be sustained. Worst of all, he might encounter the dreaded *doldrums*, a zone of the ocean in low latitudes where weeks of dull enervating calm alternated unpredictably with short bursts of squally thundery rain.

The wind patterns of the tropical belt are quite different from those associated with the mobile depressions of the mid-latitudes. Semi-permanent areas of high pressure lie about 30 to 35 degrees north and south of the equator, changing their position somewhat at various times of the year as they follow the seasonal migrations of the sun. It is this pressure pattern which dictates the tropical winds.

If the earth were not revolving on its axis, the wind would blow directly towards the equator from the two regions of high pressure. But the earth's rotation causes the air to be deflected to the west as it moves towards lower latitudes; the result is the *north-east* trade winds in the northern hemisphere, and the *south-east* trades south of the equator. The equatorial zone where the trade winds meet is called the doldrums, an area of low pressure and very light winds.

Since the air converging on the area from both sides of the equator must go somewhere, the only way it can go is up. This vast mass of ascending air is warm and humid, and a great deal of its moisture condenses as it rises, resulting in the very high rainfall associated with the squally showers of these tropical regions. A spell in the doldrums was aptly recorded by Sir Francis Drake writing in 1577: 'We drew towards the Line, where we were becalmed the space of three weeks, but yet subject to divers great Stormes, terrible Lightnings, and much Thunder.'

Words for the Wind

Weatherpeople are a pernickety lot. They like to be precise. Most of us, for example, might describe a day as windy, gusty, squally or stormy, or even blowing a gale, and regard the terms as more or less synonymous. But to people in the weather business each of these words has a very exact meaning.

To a meteorologist, the speed of the wind, unless otherwise specified, means its *average* speed measured 33 ft above the ground. The average – or *mean* – is taken over a period of ten minutes, thereby excluding from the equation those temporary ups and downs which are always present. And a gale is not just any old strong wind. A gale is said to begin when the mean speed of the wind reaches 39 m.p.h.; if the average measured over ten minutes is less than this value, there is no gale, no matter how strongly it may have been blowing for short periods in between.

Short spells of strong wind are taken care of by the two terms 'squall' and 'gust'. A gust is a sudden but very short-lived increase in wind speed where the wind rises suddenly, and then quickly reverts to near its original strength; it is a very transient phenomenon. A squall, on the other hand, is more prolonged. It is an increase in strength which lasts for several minutes before the wind decreases again to its original level.

The word *storm*, somewhat untypically for meteorologists, is used in two different ways. In one sense a storm is a deep depression of particular severity, which can be followed on the weather-map as it moves steadily along over a period of days. But the term 'storm force' is also used to describe winds of a particular strength, regardless of whether or not they are associated with any particular feature on the weather-map; they are those of Force 11 on the Beaufort Scale – stronger than 'gale' force, but less than 'hurricane' force, and blowing at a speed of between 64 and 72 m.p.h.

Coping with Heights

It is a matter of common experience that the wind seems stronger when you are high up. It is usually windier, for example, at the summit of a hill than it is in the valley below. In this particular context, of course, the difference can sometimes be explained by the shelter available down in the valley, but the principle, in general, is true: wind does increase with height. However, as one careful cynic put it: 'Nothing is absolutely true. The earth is not quite round; the sky is not quite blue; and rain is not altogether wet.' And so it is with the variation of wind with height.

Wind occurs in the first place because of differences in atmospheric pressure over the earth's surface. They bring about a movement of air from one place to another. Some two thousand feet or so above the ground, the air moves precisely as dictated by this pressure pattern, but nearer the ground it is retarded by the frictional effect of the roughness of the earth's surface. Indeed if the airflow were perfectly smooth and non-turbulent, we would expect the speed of the wind to increase uniformly with height – from near zero at the surface of the earth, to the speed dictated by the pressure pattern some hundreds of feet above.

But there is always *some* element of turbulence in the airflow. Its extent depends on a number of factors: the wind is more gusty when it is moving over rough terrain, when it is disturbed by eddies of rising air caused by solar heating, and when the wind itself is strong. Turbulence has the effect of increasing the wind speed near the ground, and this in turn decreases any difference which might otherwise exist between the wind speed at ground level and that which pertains some distance above.

Meteorologists measure the wind at a height of ten metres, or thirty-three feet, above ground level. By using this standard height, they can be sure that they are always comparing like with like when wind data from anywhere in the world are being analysed. Detailed studies have been done as to how, on average, the wind at this level compares with that measured higher up or lower down. The following relationships are generally accepted:

HEIGHT (ft)	6	15	33	80	200
RELATIONSHIP TO WIND AT 33 ft (%)	75	85	100	120	150

Beach Weather

Pleasures *al fresco* are precarious in our climate. A visit to the seaside, despite a promising start, may evoke before the day is out the moan of Keats on St Agnes Eve: 'Ah, bitter chill it was!' In summertime the beach, by its very nature, is vulnerable to the *sea-breeze.*

The irritating thing about a sea-breeze is that it occurs most readily when there is very little wind elsewhere. The pleasantly warm conditions in the garden, which one might reasonably assume should be replicated near the coast a few miles away, give way to a chilling wind, not quite cold perhaps, but intrusive enough to make one wonder why one came.

It occurs because of a developing contrast in temperature between land and sea. In the early morning there is little difference between the two but as the summer sun climbs in the sky, soil and rock quickly absorb its warmth, and heat the overlying air. Warm air tends to rise, resulting in what meteorologists call *convective currents,* columns of rising air, whose presence is often betrayed by the presence of small cumulus clouds. The actual motions of the air are complex. The net effect, however, is an upward transport of air; rather more goes *up* under the cumulus clouds than comes down in the clear spaces in between. The upward loss over the warm land is replaced by a cool current of air blowing in from the sea, drawn in horizontally across the coastline; this is the sea-breeze.

Early in the day the sea-breeze blows directly inland, at right angles to the coastline. As it becomes established during the afternoon, the tendency is for the rotation of the earth – the Coriolis Effect – to cause it gradually to change direction, so that it blows *parallel* to the coast. As the temperature contrast decreases towards evening, the sea-breeze dies away altogether.

In exceptionally warm conditions, the sea-breeze can be a welcome visitor. More often than not however, its coolness is less than pleasant. The reason for the chill is obvious when you recall that air takes its temperature from the surface over which it flows; air flowing towards you over a water surface, which in the summer months might have a temperature in the region of 14°C, must inevitably feel cool in conditions when air temperatures might otherwise be in the region of 20 to 25°C.

Romance in the Air

Foreign names have a romantic ring. As the Scottish poet Thomas Campbell put it, as long ago as 1799: 'Who hath not owned, with rapture smitten frame,/The power of grace, the magic of a name?'

Now on the scientific side of meteorology our nomenclature tends to be a little drab. It is hard to wax lyricl over hectopascals, the intertropical convergence zone, or the dry adiabatic lapse-rate. But at a more local level, weatherpeople have inherited a treasure-house of romantic terminology. In many parts of the world winds with well-known characteristics blow regularly from a certain quarter at particular times of the year; they are such familiar visitors that they have been given their own special names.

Perhaps the best known local wind is the *Mistral.* Bitterly cold, dry and violent, for long periods during the winter and early spring it sweeps down the Rhône Valley to the Côte d'Azur, forcing everyone to seek refuge indoors. Another unpleasant but regular visitor to those parts is the *Bora* – also cold and dry – which blows down to the Adriatic from the mountains of Albania and what used to be Yugoslavia. Across the Pyrenees, Spain has the cool *levanter* and the warm *Leveche.*

On the other side of the Mediterranean the *Sirocco* and the *Khamsin* are both hot, dry and dusty winds, blowing northwards from the Sahara. The Khamsin in particular is often so hazy with fine dust as it flow across Egypt, that artificial light is required at mid-day.

Across the Atlantic the US has its own indigenous nomenclature. The *Chinook* is a dry warm wind blowing from a westerly direction down the eastern slopes of the Rocky mountains into Wyoming and Montana. It sets in very suddenly, and sometimes raises the air temperature by as much as 20°C in less than half an hour, causing a rapid melting of snow, hence the other name by which it is sometimes known: the 'Snow-eater'.

An easterly airflow, on the other hand, causes the hot, dry *Santa Ana,* which sweeps down the deep canyons of southern California towards the Pacific coast, heavily laden with particles of piercing dust. And much further south again, in South America, the hot dry westerly wind which flows down from the Andes onto the plains of western Argentina, is known locally as the *Zonda.*

The Invisible Enemy

Admiral Lord Collingwood is best remembered for his stirring exhortation to his troops on the eve of the Battle of Trafalgar: 'Now, gentlemen, let us do something today of which the world may talk hereafter.' But the Admiral was involved in another incident three years later which must, in a personal sense, have been almost as memorable.

He was aboard his flagship *Ocean* at the blockade of Toulon in December 1808, when, according to an eye witness:

> At that moment the *Ocean* was struck by a very heavy sea, which threw her nearly on her beam-ends – so much so that several of our men called out 'The Admiral's gone down!' But in a few seconds I had the pleasure to see her right herself again. We understood afterwards that the blow completely disabled her; it was the most awfully terrific scene I ever beheld.

Lord Collingwood had survived bombardment at Trafalgar, but had almost succumbed to the *Mistral*.

The mistral is a cold, dry, penetrating wind from the north which sweeps down the Rhône Valley at regular intervals, most frequently in winter and early spring; it sends the locals rushing for cover, to wait behind closed shutters until the raw and dusty menace has spent itself. It is noted for the suddenness of its onset, bringing a change from virtually calm conditions to bitter gale force winds in the course of a few minutes.

This unwelcome visitor is an unusual mixture, a combination of what meteorologists call a *fall* wind and a *ravine* wind. The process starts as large amounts of air, becoming cooler through contact with the cold mountainside, slide downwards to accumulate in the valleys of the Cevennes and the French Alps. There they remain, often for some considerable time, stagnant reservoirs of cold air waiting to overflow when conditions are right.

The trigger comes when a suitable pressure pattern establishes itself, usually a depression over the Gulf of Genoa and an area of high pressure over central and northern Europe. The effect of this pressure distribution is to disturb these stagnant pools, and send a 'flash-flood' of cold air cascading down the mountains into the valley of the River Rhône below.

The 'ravine' effect is a 'funnelling' of the air which occurs as it is forced to flow southwards along the narrow river-valley. Constrictions in this narrow channel result in sudden increases in the wind speed, causing the Mistral to be strong and blustery and often dramatically destructive.

St Maury's Wind

Europe in the fourteenth century was a very unhealthy place to live. The period was remarkable for the fierce storms which caused frequent and widespread devastation throughout the countryside. It was also unusually wet and – perhaps as a result – a time of pandemic disease.

One of the most horrifying of these was *ergotism*, or St Anthony's fire, caused by the weather-related ergot blight which blackened kernels of the rye in a damp harvest. Even a tiny proportion of these poisoned grains baked in a loaf of bread would cause convulsions, hallucinations, gangrene with consequent rotting of the extremities, and ultimately death. And of course 1348 brought the great bubonic plague, the infamous 'Black Death', which wiped out half the population in many parts of continental Europe.

It was a wet and windy time in Ireland also. The major weather event of the century was a great storm which occurred on 15 January 1362, and which was documented many years ago by the late F.E. Dixon in *The Dublin Historical Record*. It was one of the worst gales ever known in historical times, and is remembered in song and story as 'St Maury's Wind'.

One contemporary chronicler described 'St Maury's Wind' thus: 'A vehement wind, which shook and threw to the ground steeples, chimney's and other higher buildings, trees beyond number and divers belfries and the bell tower of the Friars Preachers in Dublin.'

And the unwelcome event was also the subject of a long poem by one John Harding, a poem, it must be said, which must be appreciated more for its curiosity value than for its artistic merit or for the author's rhyming skill:

> In that same yere was on sainct Maury's day
> The greate winde and earthquake marvelous,
> That greatly gan the people all affraye,
> So dredfull was it then and perelous,
> Specially the wind was so boisterous,
> The stone walles, steples, houses and trees
> Were blown doune in diverse farre countrees.

Harding's reference to earthquakes, incidentally, should not be taken literally; it was common at the time for observers to assume an earthquake to be responsible for the violent vibrations often experienced during a severe gale.

The Blowing Season

Although the hurricane season in the North Atlantic lasts from late June well into October, it is mainly in late August and September that some of them survive as ghosts to haunt our European weather-maps.

A hurricane is a local manifestation of a phenomenon known generally as a *tropical revolving storm.* When they occur in the Indian Ocean, these violent storms are known as *cyclones,* in the China Seas and in most parts of the Pacific they are called *typhoons,* and near the coasts of north Australia they are known as *willy-willies.* But when they occur in the Caribbean and the North Atlantic they are *hurricanes.* The strongest winds around a hurricane are usually about twenty-five miles from the storm centre. By contrast, very close to the core, the winds are very light, and the sky is almost clear. This region – normally about fifteen miles in diameter – is called the *eye of the storm,* and can be seen very clearly as a small cloud-free area on pictures from our weather satellites.

Besides the very strong destructive winds, the chief characteristic of a hurricane is the very low atmospheric pressure at the core of the whirling circulation. The central pressure is often 50 millibars lower than that at the outskirts of the vortex, and in extreme cases the pressure difference may approach 100 millibars.

Hurricanes develop over the warm ocean just north of the equator. They usually move westwards at first, and then turn north or north-east leaving a trail of destruction behind them whenever they cross land. As it moves further north into a cooler environment, or over land where its supply of moisture is cut off, the hurricane gradually loses energy. Some fade away altogether; others retain a semblance of their identity as they are subsumed into conventional North Atlantic depressions.

By the time they reach Ireland none of them is a hurricane in the proper sense of the word, although many stormy spells are still recalled by the name of the hurricane in which they had their birth. 'Hurricane Debbie' in 1961, and 'Hurricane Charley' in 1986, are both well remembered in these islands.

Winds in a Tornado can reach 250 m.p.h. (From *HMSO Observer's Handbook*)

Hurricane Andrew entering the Gulf of Mexico on 24 August 1992. (METEOSAT image from EUMETSAT)

The Fearsome Twister

Tornadoes are very common on the great plains of the United States, and because of their concentrated power, they are probably the most feared of all weather phenomena. They strike suddenly with little warning, and can cause the most unbelievable devastation in the space of a few minutes.

Unlike the hurricane, the tornado is a very localized phenomenon, so small that it cannot be represented by a system of isobars on a normal weather-chart. It is a violent whirlwind, anything from a few yards to half a mile in diameter, and it moves forward at between 10 and 20 m.p.h., in more or less a straight line. Despite this somewhat leisurely advance, the winds raging around the swirling maelstrom reach speeds of up to 250 m.p.h. But much of the damage left behind the whirling monster is not the result of wind at all. A very sharp drop in atmospheric pressure occurs at the centre of a tornado. The fall in pressure occurs so suddenly outside buildings in the path of the vortex, that they are caused – quite literally – to explode.

With rare exceptions, tornadoes are cyclonic; they blow in an *anti-clockwise* direction in the northern hemisphere. They almost always occur in thundery weather, and because of their small dimensions, the damage is normally confined to a very narrow path along the track. And the horror is short-lived; the average life-span of a 'twister' is less than half an hour.

Tornadoes have two other notable characteristics. The first is the familiar corkscrew cloud, undulating downwards to the ground below. And the second is the loud distinctive noise of the whirlwind. People unfortunate enough to have firsthand experience of a close encounter have used strange similes to describe the sound. It resembles, they say: 'a thousand railway trains', 'the roar of a flight of jets', or perhaps most picturesque of all, 'the loud buzzing of a million bees'.

An Ill Wind

The railway bridge across the River Tay near the Scottish town of Dundee was designed and built by Sir Thomas Bouch at a cost of £670,000. It was opened with great pomp and ceremony in 1877, a triumph of nineteenth-century engineering and, it was hoped, a monument to the Age of Progress. It carried two railway lines, and had 74 spans describing a gentle curve over the two miles of its length. At its highest point there was a clear 80 ft to allow passage underneath to the largest ships of the day.

Triumph soon turned to tragedy. The Christmas season two years later was marked by a succession of very violent storms. On 28 December 1879 there occurred what was then the greatest ever railway disaster. One of the spans had become weakened by the frequent gales, and as a full train made its way across the structure in high winds the Tay Bridge collapsed; the train and its passengers plunged headlong into the icy waters of the River Tay, and seventy-five people were drowned.

That disaster had important implications for the young science of meteorology. There were claims and counter allegations, not only about the strength and construction of the bridge, but also about the relative merits of the available instruments for measuring the strength of the wind. As a consequence the Royal Meteorological Society set up a Wind Force Committee to 'investigate and report on the best mode available for a satisfactory solution to the entire question of wind's force.'

The most widely used instrument at the time was the Robinson cup anemometer, but there was disagreement among scientists of the day about its accuracy. Many were convinced that it over-registered the average speed of the wind, and under-registered the gusts (because of the mechanical lag associated with the momentum of the whirling cups). It was feared that the available statistics on high winds were therefore unreliable, with obvious consequences for the construction of wind-sensitive structures like the Tay Bridge.

The Committee allocated to a Mr William Henry Dines the magnificent sum of £6 to investigate the question, and he diligently did so in the back garden of his house in Hersham. His experiments confirmed the doubts which had been expressed about the errors inherent in the current Robinson design, and led to the development of the Dines pressure tube anemometer, which became the standard wind-measuring instrument for over three-quarters of a century.

Dividing the Waters

Had accurate shipping forecasts been available in Antonio's day, he might have been saved a great deal of worry, and the world would have been much the poorer for never having heard of *The Merchant of Venice*.

Antonio, if you remember, had 'an argosy bound to Tripolis, another to the Indies, a third at Mexico, and a fourth for England'. Although his risk was well spread, he was vulnerable to storm and tempest, and Shylock was quick to take advantage of his reported misfortune in this respect. Shipping forecasts were still to come. They began in 1861, after a number of spectacular shipwrecks had highlighted the benefits that could accrue from advance warnings of severe weather.

Admiral FitzRoy, asked to do something about the problem, was soon in a position to issue daily wind forecasts for sailors. For this purpose he divided the coasts of Britain, Ireland and the north of France into seven 'districts', and published a two-day 'opinion' for each district. Shipping forecasts are still with us, but have changed a great deal over the years in terms of both accuracy and presentation.

The 'districts' are gone – or at least changed. Nowadays the shipping forecasts we hear on Irish radio refer to a strip of water about 35 miles wide around the coast of Ireland. This strip is sub-divided as necessary by referring to headlands around the coast, thus identifying in detail the zone to which the forecast refers.

The arrangement in Britain is somewhat different and less obvious. By long-standing international agreement, the British authorities have a formal responsibility to provide weather information to shipping far out into the Atlantic. They divide this zone of responsibility into twenty-eight 'sea areas' of varying sizes, and forecasts are provided for each of them in a clockwise order around the islands of Britain and Ireland – starting with 'Viking' near Norway, and ending with the area southeast of Iceland.

Most of the sea areas are named after some geographical landmark in the locality: 'Lundy', 'Fair Isle', 'Faeroes', and 'Rockall', are all called after islands they contain. Those along the east coast of England and Scotland are named from rivers flowing into them: 'Thames', 'Forth', 'Tyne' and so on; 'Shannon' too, comes into this category. And yet others are named after headlands or lighthouses, like 'Fasnet', 'Portland' and 'Finisterre'.

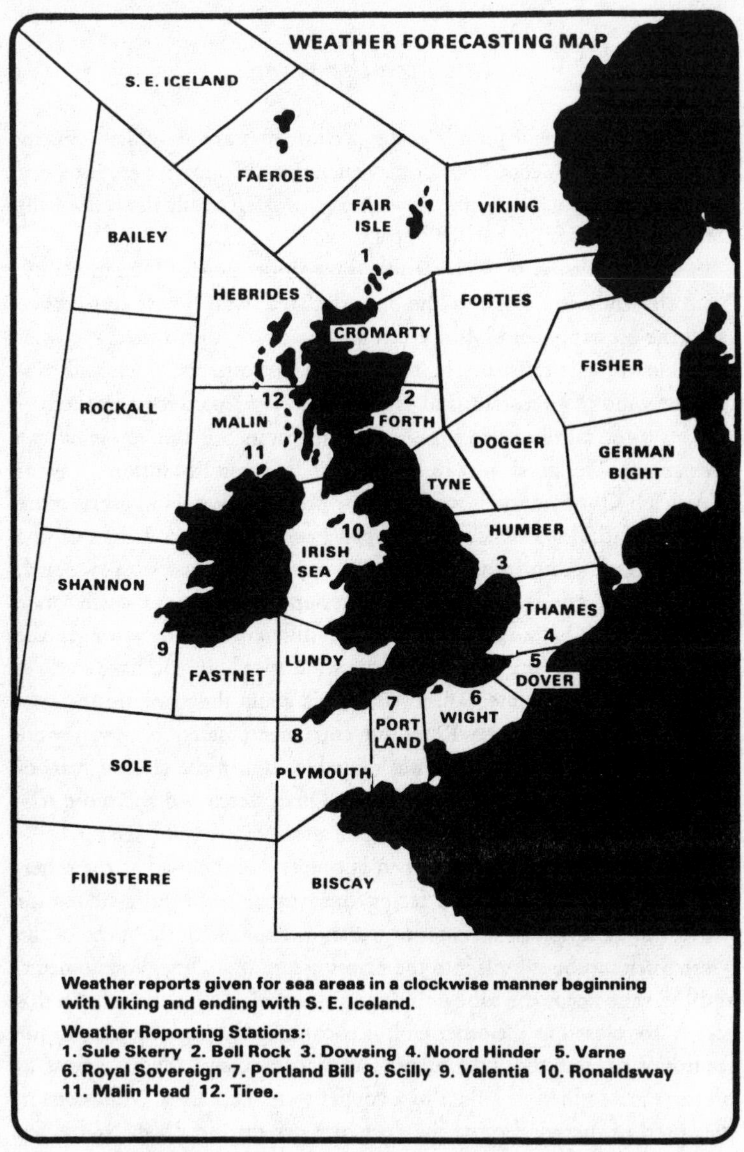

The sea areas used in British shipping forecasts. (*UKMO Annual Report*, 1986)

To Catch the Wind

The English dramatist John Webster, a contemporary of William Shakespeare, was the Vincent Price of his generation. His plays were intended to shock, and shock they did; *The Duchess of Malfi* rivals the best of the Hammer classics in its horrific impact.

A theme beloved of Webster was the pointlessness of human vanity. 'Vain the ambition of kings', he once declared, who 'weave but nets to catch the blowing wind.' And yet in a strange way, this is precisely what the meteorologist must do; he must devise instruments which will capture the wind of an instant, and provide from it a record for posterity.

The most familiar wind-measuring instrument is the *Robinson cup anemometer.* Designed in 1846 by Thomas Romney Robinson, Director of Armagh Observatory, it works on the principle that wind exerts more pressure on the *concave* side of an open hemisphere than it does on the *convex* surface. A horizontal bar, free to rotate and fitted with two such hemispheres – one at each end, facing in opposite directions – will rotate with the wind because of this pressure differential. The stronger the wind, the faster it whirls; to record the wind speed, all you have to do is attach the device to a mechanism which will count the revolutions.

Earlier versions of the Robinson cup anemometer, however, were found to be less accurate than was desirable, and in the closing years of the nineteenth century, William Henry Dines developed the more reliable instrument, the Dines pressure tube anemometer (see page 53).

Imagine, first of all, a tube open at one end and closed at the other, and held with the open end facing the wind; the pressure of the air inside will be raised by an amount which increases with the speed of the wind. Such a tube is built into the pivoting arm of a Dines anemometer, and the vane keeps the tube pointing into the wind. But the tube in this case is not closed at the other end; it is connected by a long pipe to the recording part of the instrument housed below, and the variations in pressure are applied to a float in a copper water-container. Variations in the speed of the wind cause the float to move up and down, and a pen attached to it provides a trace which closely corresponds to the speed of the wind.

For many years the Dines pressure tube anemometer was the standard instrument for recording wind speed. In recent years however, the design of the Robinson has been much improved, and it now rivals its one-time successor in the accuracy of its wind values.

IV *Blowing Hot and Cold*

An Elementary Rhythm

At a very early stage in his or her career each would-be weatherperson learns 'doctrines fashion'd to the varying hour' – the mysteries of the *diurnal cycle* are unveiled. The many individual tunes, which orchestrated as a chorus make up the symphony we know as 'weather', show little sign of harmony when taken in isolation; parameters like temperature, wind and rainfall change in a disorganized way from hour to hour and from day to day. But if a series of measurements is averaged over a long period, a pattern is seen, a daily rhythm rooted in the motion of the sun.

The temperature of the air, for example, is at its lowest – on average – shortly after dawn. Air takes its temperature from the ground beneath, and on a clear night the ground gets colder and colder until the rising sun begins to replace some of its lost heat. But for some time after sunrise, the earth is still colder than the air above it so the temperature of the air continues to fall. Only when the ground becomes *warmer* than the air, does the air temperature begin to rise. Hence the lowest air temperature is not *at* sunrise, but an hour or two *after* it. And in a similar way, the highest temperature of the day occurs – on average – an hour or two after noon.

This temperature cycle affects the other weather elements. Warmth near the surface is conducive to rising currents of air in the atmosphere, which in turn breed clouds. The afternoon, therefore, tends to be the cloudiest part of the day. And clouds often produce rain, with the result that rainfall, too, peaks in the afternoon.

If the pressure pattern is steady, the wind waxes and wanes in much the same way; it is at its lightest around dawn, and strongest on average in the afternoon. Relative humidity, on the other hand, varies in the opposite direction. You may remember that the *lower* the temperature, the *higher* the relative humidity for a given amount of moisture in the air; relative humidity, therefore, peaks just after sunrise, when the temperature is at a minimum, and is lowest when the sun has just passed its zenith.

Most of these daily ups and downs depend, directly or indirectly, on the transient temperature of the earth's surface. Land areas react rapidly to the effects of the sun, but the temperature of the sea, by contrast, varies little throughout the twenty-four hours. For this reason, the diurnal variation of nearly all meteorological elements is a good deal less pronounced near the coast than it is in inland areas.

Meteorological Extremism

Superlatives have an abiding fascination. On any particular day two temperature values stand out as being by far the most interesting: the *highest* and the *lowest*, or, as they are more usually called, the *maximum* and *minimum* temperatures. One way of finding, for example, the highest temperature, would be for the observer to keep a constant eye on the thermometer, making sure to spot the exact instant when it hits its highest point. But the exercise would be tedious and time-consuming; as I believe somebody once said, 'There is a better way!'

A thermograph is one solution. Thermographs provide a continuous record of the temperature on a paper chart, and most of them work on the principle that metal expands when it is heated and contracts again as it cools. The activating mechanism is a metal spiral which coils and uncoils with the variations in temperature; this action is used to control the movement of a recording pen. From the rising and falling trace provided by the thermograph, the maximum and minimum temperature to occur in any given period can easily be deduced.

More accurate values, however, can be obtained by using special thermometers. The *maximum thermometer* uses mercury, and is very similar to that used for measuring the routine hourly values of temperature. But it has an important peculiarity: near the bulb of the thermometer is a narrow constriction past which the mercury can flow only with considerable difficulty. As the temperature rises, the thermal expansion of the mercury in the bulb is sufficient to force it through the narrow opening until the highest value is reached. But when the temperature begins to fall again, there is no pressure on the fluid to force it back past the constriction; the mercury column breaks, and the instrument continues to register the value of the highest temperature.

The minimum thermometer works on a quite different principle. It contains alcohol instead of mercury, and immersed in the fluid is a very light dumb-bell shaped object called an 'index', which is free to slide along the thermometer tube. The thermometer is set each evening, mounted horizontally with the index inside the column of alcohol and just touching the end of it. As the temperature falls and the spirit contracts towards the bulb, the fluid drags the index with it until the lowest value is reached during the night. But as the temperature rises again, the expanding column of alcohol leaves the index behind at the lowest point, providing a 'marker' to identify the minimum temperature.

Eccentric Spells

Alexander Buchan was a Scotsman. In 1869, having spent much time examining weather records and making his own weather observations, Buchan asserted that at certain regular times of the year the weather was consistently either warmer or colder than the calendar would suggest it ought to be. Buchan himself never claimed that his spells were an infallible guide to the weather, and indeed only suggested that they seemed to apply in his native part of Scotland. But his ideas caught the popular imagination, and Buchan Spells were widely accepted for a time throughout these islands as an established feature of our weather.

Buchan was no eccentric dilettante. He was in fact one of the best-known meteorologists of his day, and was for more than forty years the presiding genius of the Scottish Meteorological Society. He identified, in all, six cold periods and three unseasonably warm ones. The warm spells were 12–15 July, 12–15 August and 3–14 December; the cold spells were 7–14 February, 11–14 April, 9–14 May, 29 June–4 July, 6–11 August, and 6–13 November.

In 1927 Buchan's theories became a matter of some controversy. At that time, the Westminster Parliament was discussing the possibility of establishing a fixed date for Easter, on the first Sunday after the second Saturday in April; this would have placed Easter each year between the 9th and 15th of April. But it was noticed that these dates closely corresponded with Buchan's cold spell of 11–14 April. The chilling prospect of cold Easters in perpetuity was too much, and the idea was dropped.

Unseasonably warm or cold periods do, of course, occur in most years but not with a regularity which would make them of any value for predicting the weather. Irish tradition, for example, identifies a regular cold snap in early May called *Scairbhin na gCuac*, which is close to Buchan's cold spell in the same month. And another common feature of the weather in Ireland is the occurrence of a few mild days in early December, to correspond with Buchan's period of the 3rd to the 14th. But Buchan Spells do not stand up to rigorous climatological examination. Life – and the weather – are a good deal less predictable than that.

Alexander Buchan (1829–1907), originator of the 'Buchan Spells'.
(Courtesy of Royal Meteorological Society)

Think of a Number

When Samuel Pepys arose on the morning of 27 November 1662 there was an unfamiliar chill in the air. He recorded the experience in his famous Diary: 'At my waking, I found the tops of the houses covered with snow, which is a rare sight, which I have not seen these three years.'

Pepys could only observe the coldness in a subjective way. At the time there was no accepted method of assigning a number to temperature in the way we do today. The problem was highlighted by Corkman Richard Boyle just three years later, when he wrote in 1665: 'We are greatly at a loss for a standard whereby to measure cold. The common instruments show us no more than the relative coldness of the air, but leave us in the dark as to the positive degree thereof; whence we cannot communicate any idea thereof to any other person.'

But the matter was receiving attention. Indeed between the years 1641 and 1780 at least thirty-five different scales of temperature were proposed and in limited use. The number of candidates for immortality was eventually reduced down to three: Reaumur, Fahrenheit and Celsius. And it was not until the present century that the Celsius temperature scale emerged as the outright winner.

Anders Celsius was a Swede, born at Upsala on 27 November 1701. In due course he became Professor of Astronomy at Upsala University, and it was on 15 January 1742 that he unveiled his bright idea in a paper to the Swedish Academy of Sciences entitled 'Observations About Two Fixed Degrees on a Thermometer'. He demonstrated the unsatisfactory nature of the existing temperature scales, and went on to present one of his own, with 100 degrees between the boiling-point and freezing-point of water.

Celsius's scale had one very surprising feature: it was the reverse of the centigrade scale we know today. He designated 100° as the *freezing-point* of water, and zero as its boiling-point, and justified its topsy-turvy nature on the grounds that it avoided the use of the minus sign in winter. The scale subsequently was turned right-side-up by his colleague and compatriot Carl Linnaeus. So, to be strictly correct, we should designate temperature today in 'degrees Linnaeus' rather than in 'degrees Celsius'.

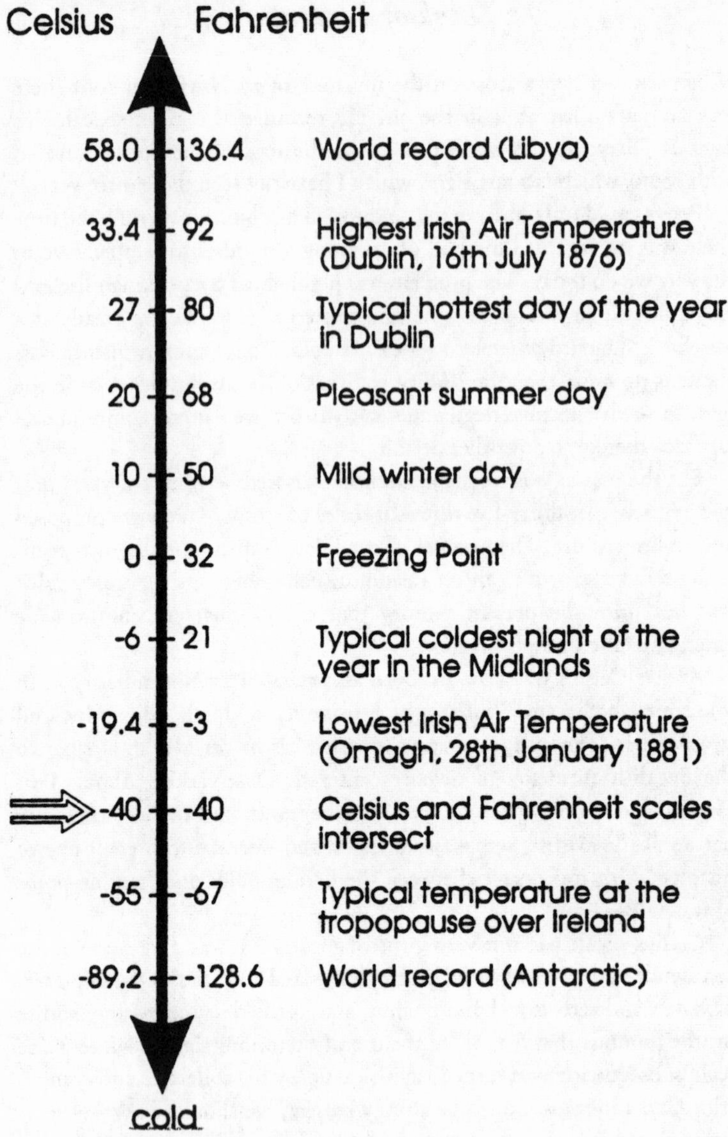

The Fahrenheit and Celsius scales compared. (Anne McWilliams)

City Heat

'No city', according to Cyril Connolly, 'should be too large for a man to walk out of in a morning.' But most cities are, and living in a large city demands a change in lifestyle for the rural immigrant; however it is less well known that it also involves a change – albeit small – in climate. On a clear night, in the centre of a densely populated built-up area, the temperature may be five or six degrees higher than that of the surrounding countryside.

The *heat island* phenomenon, as it is called, is now a well-known feature of all major cities. It is most obvious in the late evening and early at night, and it reaches a maximum about two or three hours after sunset. Many factors contribute to its existence; some of them are obvious, but others are less so.

In the first place, buildings which are artificially heated in winter-time to a temperature which suits our human comfort, lose heat continually to a colder outside environment. If this happens on a large scale, the air temperature outside will increase. Secondly, the brick and concrete which are present in abundance in urban areas are good absorbers of heat; they store the warmth received from the sun during the day, and release it slowly to the surrounding atmosphere at night, like giant storage heaters. The compacted soils beneath roads and parking areas act in the same way, quite unlike the relatively loose and 'air-filled' soil of agricultural land

Another factor which contributes to the urban temperature anomaly is the fact that tall buildings often block the wind which would otherwise disperse any pockets of warmer air. And finally, the rapid drainage of surface water through the shores and gullies of the city streets means that less energy is dissipated by evaporation than is the case in a rural setting, where the water stays near the surface trapped in the sponge-like soil. The energy used for evaporation reduces the temperature over the countryside, but no equivalent drop takes place in the urban environment.

Heat islands near large centres of population are easily detectable by the very accurate temperature sensors of today's weather satellites. The anomaly is most common on calm clear nights, and is a good deal less noticeable in dull windy weather. It is also a very localized phenomenon; the boundary of the heat island is quite pronounced, and the temperature drops rapidly to the surrounding norm at the edge of the built-up area, reinforcing the traditional contrast between town and country.

Blanket Economy

Blankets conserve heat. Or, as Dean Jonathan Swift wryly remarked all those years ago: 'Tis very warm weather when one's in bed.' But the principle is not confined to the somnolent atmosphere of the domestic bed-chamber. It has a much wider application; our whole planet has a blanket to keep it warm, and our fuel bills in winter-time would be considerably higher were it not for the frequent presence of a protective layer of cloud.

Clouds have a very marked effect on air temperature after dark, particularly during long winter nights if the wind is light. In the absence of cloud, the earth loses heat by long-wave radiation. The energy is radiated out into space, causing the earth's temperature, and hence the temperature of the air in contact with it, to fall. The cooling process continues right through the night, with temperatures reaching a minimum around dawn. As a result the thermostat controlling the domestic central heating system stays in the 'on' position, and fuel is used virtually continuously.

To a large extent the presence of a layer of cloud prevents this heat loss. The earth continues to radiate energy as usual, but the outgoing radiation is absorbed by the clouds a short distance above, and the lost heat is prevented from escaping out to space. The clouds in turn re-radiate heat downwards, and this downward radiation may often be enough to balance the heat-loss from the ground. The net result is that the minimum temperature in overcast conditions may be as much as ten degrees higher than that on an otherwise similar but cloud-free night.

During cloudy conditions in winter-time there is little difference between night-time and daytime air temperatures. The cloud, of course, also prevents daytime temperatures from rising as high as they otherwise might, by reflecting the incoming solar radiation and preventing it from reaching the earth. This though, makes less difference to the daily temperature cycle than does the night-time effect, because winter days are relatively short.

From an economic viewpoint, therefore, clouds are welcome visitors in winter-time. A clear starry night with a full moon is a pleasure to look at, but we pay a price in terms of larger fuel bills, as central heating systems try to cope with the freezing temperatures which accompany the clear skies.

Shades to Shiver By

In the strictest sense, frost is said to occur when the temperature – wherever it may be measured – falls to zero degrees Celsius or below. Meteorologists, however, with their usual pedantry, distinguish between *air frost* and *ground frost.*

The standard height for measuring the temperature of the air is four feet above the ground. The temperature is also routinely monitored at ground level, and the two values are often very different; on a cold clear night, the temperature on the grass may be several degrees lower than that a few feet above. *Ground frost,* therefore, is a much more frequent visitor than air frost, and occurs when the temperature *on the ground* is zero or below.

Either ground frost or air frost may occur with no visible signs of their presence – a so-called *black frost.* The only way of detecting them with confidence is with the aid of a thermometer. But the term 'frost' is also used to describe the white icy deposits which are common in freezing conditions, deposits which can be seen and recognized for what they are without recourse to a thermometer or to any other instrument.

White frost, or *hoar* frost, is a white crystalline deposit often seen on grass and other outdoor surfaces on a bright clear winter morning. Indeed the term 'hoar' is an Old English word for 'white'. It is closely related to dew; dew occurs when water vapour in the air condenses onto a cool surface underneath, but if the temperature of the underlying surface is below zero, the deposition is in the form of soft, white ice-crystals – giving hoar frost. Indeed on some occasions, there may already be a deposit of dew before the temperature falls below zero, and when the dew subsequently freezes, it is known as *white dew,* or *silver frost.*

From many points of view the most troublesome and dangerous form of frost is *glazed frost.* It occurs when rain falling from warmer air above freezes on striking the cold ground, and forms a clear transparent sheet. The result is similar to that which occurs when a layer of water already on the surface of a road freezes to produce the treacherous conditions popularly known as *black ice.*

Frost Fighting

Frost, in general, is not the gardener's friend. Horticulturally troublesome frost comes in two guises. *Wind frosts* are normally associated with freezing conditions over northern Europe, and as cold easterly winds sweep in over Ireland, vulnerable plants suffer a blackening or a browning of their leaves; in very severe cases, the plant itself may not survive an episode. *Radiation frost* is much more common; it occurs on calm, still, starry nights, as the ground cools rapidly by losing heat in the form of long-wave radiation out to space.

There are ways in which frost damage can be minimized. Perhaps the most effective, insofar as it may be possible, is careful selection of the planting site to avoid 'frost hollows', depressions in which pools of cold dense air accumulate, having drained downhill along the surrounding slopes. Indeed an artificial frost hollow can be created inadvertently if, for example, a shelter-belt of thick foliage on sloping ground acts like a dam to create a reservoir of cold air on its uphill side.

Frost damage to low-growing plants can be contained by protecting them with straw, or glass, or plastic sheeting; the covering retards loss of heat by radiation, and slows the rate of fall of temperature beneath. Less obvious is the technique of *advance irrigation*: wetting the ground the day before the frost occurs; it depends for its effectiveness on the fact that wet compact soil absorbs significantly more energy during the day than it does if it is dry and loose, and this extra store of heat stands it in good stead during the long cold night to follow.

Perhaps the most surprising way to protect your plants from frost is to sprinkle them continuously with cold water. The effectiveness of this method lies in the fact that water, as it freezes, releases heat, the so-called *latent heat of fusion*; as the drops of water freeze upon the leaves, the heat released forestalls the damage to the foliage. Unfortunately the technique has a severe disadvantage in the average unautomated garden; the plants must be re-wetted every thirty seconds while the freezing lasts, because any interruption of the sprinkling results in damage even more severe than if no preventive measures had been taken in the first place.

A Chill in the Air

Temperature alone does not measure well 'the icy fang and churlish chiding of the winter's wind'. As we all know from personal experience, the stronger the wind, the colder it feels, and very cold weather always brings talk of *wind-chill*.

The *wind-chill equivalent temperature* – or WET for short – is not an actual temperature; it is a measure of the influence of the wind on the *perceived* temperature. It is in a sense a measure of discomfort; it answers the question: 'How cold would it have to be if there were *no* wind blowing, for me to feel as uncomfortable as I do now?'

Meteorologists are a little suspicious of wind-chill. Their main objection is that the concept may be misapplied. The idea is useful in the case of humans or animals in situations where they are likely to suffer adverse effects from exposure, or in studying the problem of heat-loss from buildings during the winter-time. Indeed wind-chill is a useful concept in any situation where *warm* objects are exposed simultaneously to both wind and low temperatures. Put simply, in such a situation the cold wind carries the heat away.

But it is quite wrong to apply the concept to unheated inanimate objects, who, so to speak, have no heat to lose. The temperature of a stationary motor car engine, of an oil storage tank, or of growing plants will not drop below the temperature of the surrounding air no matter how strongly the wind blows.

And even in the case of human beings, the wind-chill equivalent temperature quoted on the media should not be taken too literally. The extent to which we feel the cold depends on many other factors besides wind and temperature: it depends on how much clothing we have on, on our age and physical condition, on whether we are physically active at the time or not, and indeed on whether the wind comes from behind or blows directly onto the face.

Used sensibly and appropriately though, the concept of wind-chill can be a useful one. Its popularity stems largely from the fact that at a superficial level it is very easy to understand. We have all felt chilled by a stiff breeze when standing in the cold, and it is nice to be able to put a number on the extent of our discomfort.

Macabre Extremes

Let's talk of graves, of worms, and epitaphs;
Let's sit upon the ground,
And tell sad stories of the death of Kings.

Or maybe not! Perhaps, on second thoughts, we should take it bit by bit; let's merely talk of graves!

The grave, without doubt, is a damp inhospitable spot. But, contrary to the accepted wisdom of Victorian melodrama, it is not a particularly cold place to be. At times, indeed, it may be warmer six feet down than here on top.

Temperature variations, as a general rule, are greatest at ground level. The surface of the earth responds quickly to heating by the sun, and loses energy rapidly in its absence. Beneath the ground though, the soil reacts slowly to the temperature changes overhead. The deeper you go, the less the temperature varies.

At a depth of 4 in. below the surface, the difference between the average temperature of the warmest month and the average temperature of the coldest month is about 13°C. At a depth of 4 ft this difference decreases to about 7°C, and 30 or 40 ft below ground, there is little variation at all in the average temperature throughout the year. It remains more or less constant.

The relative stability of temperatures underground arises because the layers of soil above act like blankets, making it difficult for the heat of the sun to penetrate, but also retarding the loss of heat on cold winter nights. In the summertime, therefore, the temperature near the surface is higher than that lower down; in winter, cooling is most marked at the surface, and the temperatures underneath are quite a few degrees higher.

The *diurnal variation* of temperature also decreases with depth. Air temperature rises during the day to reach a maximum in the early afternoon, and drops to a minimum around dawn. This pattern, albeit less marked, is detectable beneath the ground to a depth of about 4 ft, but below that the temperature remains more or less constant throughout the twenty-four hours.

And not too far beneath the surface, the extremes of a hard winter are hardly felt at all. Sub-zero temperatures occur below 8 in. only during a prolonged cold spell, and temperatures below zero at depths greater than a foot have never been observed in Ireland.

High-Low Temperatures

It is common knowledge that temperature normally decreases with height. By taking frequent measurements at various points on the slopes of high mountains, it was known for a long time that the average rate of decrease was of the order of 2°C for every 1000 ft. Until the beginning of the present century, it was assumed that this decrease continued indefinitely, until somewhere out in space the temperature reached what scientists call 'absolute zero', the lowest temperature theoretically achievable, which turns out to be about –273°C.

But it was not so. In 1902 a French meteorologist called Teisserenc de Bort made what has since been called 'the most striking discovery in the whole history of meteorology'. He found, by using recording instruments carried aloft by free balloons, that above a certain height the temperature stops falling and may even begin to rise again.

De Bort decided that the atmosphere has two layers. Near the surface of the earth is a turbulent lower layer containing clouds, rain and storms, and all the familiar ingredients of our changeable weather; he called it the *troposphere*, meaning the 'sphere of change'. Above it is an upper region of thin, almost cloudless air which he called the *stratosphere.* The boundary between the two, the level at which temperature ceases to fall, he called the *tropopause*, 'the end of change'.

This change in the rate of fall of temperature occurs very abruptly, so that the tropopause is extremely well defined on diagrams which show the variation of temperature with height over a particular spot; diagrams known to meteorologists as *tephigrams.* The tropopause surface slopes from the equator towards the poles; over the North and South Poles it is about five miles high, while in equatorial regions temperature continues to fall with height up to a distance of eleven miles or more above the earth. Over Ireland the average height of the tropopause is about eight miles, and the temperature at that level averages about –55°C.

In the stratosphere, temperature remains almost constant with height, or may even increase slightly. The relatively high temperature in these regions is explained by the presence of ozone, which is a very effective absorber of the ultraviolet radiation emitted by the sun. The process of absorption not only increases temperatures in the stratosphere, but also prevents much of the potentially harmful radiation from reaching the earth below.

A Toe in the Water

Land masses in the northern hemisphere are at their coldest in January and at their warmest in mid-July, Ireland being no exception. But the sea lags behind; it reaches its extreme values of temperature about a month later in each case, being coldest in February and warmest in late August. The increase in temperature which takes place from early spring to late summer shows none of the daily ups and downs that characterize the temperature of the air; the change is slow and gradual, but relatively steady.

Statistics about the temperature of the sea are gathered by ships, who take the trouble to measure it at regular intervals. Their technique is disarmingly straightforward; they simply heave a bucket over the side, haul it up full of water, and take the water-temperature with a thermometer. In recent years, modern technology has provided a more efficient method of obtaining this information. Sensors on board orbiting satellites react to the long-wave radiation emitted by the surface of the ocean, which is in turn dependent on the temperature of the water. By this means, maps showing the variations in sea surface temperature from place to place can be obtained almost instantaneously.

The warmest water in the vicinity of Ireland is usually to be found in the south-west, near the coasts of Kerry and Cork; the sea is coldest, on the other hand, off the coast of Antrim. The temperature difference between these two zones is usually about 2 or 3°C. Mid-winter sea temperatures off the south-west coast are normally a little less than 10°C, while in the north Irish Sea they drop to around 7°C. At the other extreme, the summertime maximum in August reaches 15.5°C along most of the south coast, but a mere 13°C between Fair Head and the coast of Scotland.

These, of course, are *average* values. In any particular year the *actual* sea temperature may differ from the average by as much as a degree or two. And these figures also represent the temperature some distance from the coast; close to shore, and particularly in long inlets, the summer sea might be expected to be slightly warmer.

A Campbell-Stokes Sunshine Recorder. (From *HMSO Observer's Handbook*)

A Precious Commodity

Two factors affect the daily sunshine record. The amount of cloud in the sky is obviously important, but so too is the length of the solar day. At the two equinoxes – around 21 March and 22 September – the sun rises in the east, sets in the west, and the day is exactly twelve hours long. In December, however, it describes a much smaller arc, rising in the south-east and setting only seven or eight hours later in the south-west, staying low in the sky all the time. But in mid-summer, the sun climbs to near 60 degrees above the southern horizon, having risen in the north-east and travelled right around to the north-west in the course of a day a full seventeen hours long.

Because of this longer period of daylight, there will obviously be more sunshine on a clear day in June than on an equally clear day in December. Strictly speaking, therefore, when we compare the sunshine figures for the two months, we are not comparing like with like; June has an unfair advantage.

Sunshine is recorded by meteorologists using the Campbell-Stokes Sunshine Recorder – an elegantly simple instrument consisting of nothing more complicated than a solid glass sphere and a thin strip of card. When the sphere is placed in bright sunlight, the parallel rays coming from the sun are focused to a point as they pass through it. The card is placed at the 'hot spot' where the rays converge, and develops a scorch-mark, a principle familiar to every schoolboy who has experimented with a magnifying glass on a sunny day.

This simple and effective method of measuring the duration of sunshine was first mooted in Rome around 1646 by one Athanasius Kircher; it was subsequently taken up by a Scotsman, John Francis Campbell, and later improved upon in the late nineteenth century by an Irish scientist, Sir George Stokes. As the sun moves across a clear sky, a continuous mark is burned onto the card; alternatively, if the sun is obscured by cloud, no mark at all appears. To obtain a perfect record of the day's sunshine, all that is necessary is to clamp the card in a semicircular guide at the correct distance from the sphere.

V *The Vapours*

Drying Out

'Progress', according to the American writer e.e. cummings, 'is a comfortable disease.' For most of us, the problem of keeping warm indoors in winter has been solved. Central heating is commonplace, both at home and in the workplace, and indoor temperatures can easily be kept at a comfortable level.

The concept of central heating is not new. The Romans used it in their villas two thousand years ago, but their techniques were largely forgotten during the chaos of the Dark Ages, and for many centuries people kept warm during the winter by coal or wood fires, variously positioned in all shapes and sizes of grates and stoves. But although central heating solves the problem of warmth, it can have an adverse effect on another important aspect of our comfort. When we control temperature we inadvertently change the humidity.

Relative humidity is temperature dependent; roughly speaking, it is the ratio between the actual amount of moisture present in a body of air, and the amount of moisture the air is *capable* of containing at the same temperature, the amount of moisture it would contain if it were saturated. As we know, warm air can hold more moisture than cold air.

Suppose, for example, the outside temperature is 10°C and the relative humidity is 80 per cent. At 10°C, air is capable of holding about 9 grams of water vapour per cubic metre, but since in this case the relative humidity is only 80 per cent, each cubic metre in fact contains only about 7 grams.

Now if the central heating inside the house raises the temperature of this air to 20°C, since no moisture has been added it still contains only 7 grams of water vapour per cubic metre. But at 20°C air each cubic metre of air can absorb 17 grams of water before becoming saturated, so at the higher temperature the relative humidity is only around 40 per cent. The air feels dry, sometimes even to the extent that we are uncomfortable, with stinging eyes and irritated skin.

The problem of very low humidity indoors is most likely to arise when the air coming into the house is dry to start with, which happens most frequently in the cold showery north-westerly airflow behind a cold front. And somewhat paradoxically, it is a winter problem; in the summertime, with windows and doors open for much of the time, the humidity indoors is normally similar to that outside.

Attacking the Vapours

In one of his more dismal moods, Hamlet felt the air around him to be 'a foul and pestilent congregation of vapours'. This was not an original view of things; in very olden times all weather phenomena were believed to result from the interaction in the atmosphere of two kinds of 'vapour', exuded by the earth under the influence of the sun's heat. According to the French philosopher Guillaume Du Bartas, the sun,

> Two sorts of vapours by his heat exhales
> From floating deeps, and from the flowery dales;
> The one is somewhat hot, but heavy, moist and thick,
> The other light, dry, burning, pure and quick.

Modern meteorologists have simplified the problem somewhat. They concentrate on a single vapour: *water vapour,* the amount of which in the air at any particular time determines the humidity.

They have devised many ingenious ways of measuring it. The simplest method is to use a substance which changes its physical characteristics as it absorbs or loses moisture. Human hair has this property – it expands or contracts – and it is often used to construct instruments to provide a continuous record of the relative humidity of the atmosphere. Sometimes *goldbeaters' skin* is used in the same way.

The most accurate way of measuring humidity is to use a *psychrometer.* This consists of two thermometers side by side, one of which – the 'wet bulb' – has its bulb enclosed in a 'glove' of muslin dampened by distilled water. Evaporation from the muslin results in a drop in the temperature of the wet bulb; the dryer the surrounding air, the greater the rate of evaporation, and the greater the drop in temperature. The difference between the readings of the two thermometers is thus a measure of the moisture content of the air.

And there are even cleverer ways. The *Dew-Point Hygrometer,* for example, directs a thin pencil of light into a very sensitive photoelectric recorder by way of a reflecting mirror. Obviously if condensation occurs on the mirror the amount of light reaching the recorder will be dramatically reduced. To measure humidity, the temperature of the mirror is electrically controlled, and brought to the point – as judged by the photo-electric recorder – where condensation has just begun to form. The temperature of the mirror is then precisely the dew point of the surrounding air, and from this the humidity can be easily calculated.

A Life-Giving Cycle

The torture to be endured by Coleridge's ancient mariner, if you remember, was that there was: 'Water, water, every where,/Nor any drop to drink.'

Although there is a great mass of water on the earth, most of it is heavily laden with salt and is contained in the world's oceans. As far as humans are concerned, the most vital water supplies are the relatively small quantities contained in rivers, lakes, ground water, and the water that circulates between the atmosphere and the continents. The husbanding of this water is the science of hydrology, and it is achieved through analysis of the *hydrological cycle.*

There are altogether some 1500 million cubic kilometres of water on the planet; enough to fill a giant cubic tank with sides 700 miles long. About 97 per cent of it resides in the oceans, and most of the remainder is locked semi-permanently in ice sheets and glaciers; the fresh water resources used by humans, and the water used in atmospheric processes, are only a tiny proportion of the whole. Only about 14,000 cubic kilometres are present in the atmosphere at any given time in the form of clouds or invisible water vapour, and it is estimated that if all this were condensed, it would yield a layer of water no more than 3 centimetres deep.

The driving force behind the hydrological cycle is the radiant energy of the sun. Heating of the sea causes *evaporation,* by means of which water is changed from the liquid into the gaseous state, and absorbed as part of our atmosphere. In due course, this water vapour falls back to earth again as *precipitation,* mainly in the form of rain or snow. Some of it evaporates from the surface; much of it infiltrates the soil, percolating downwards until it reaches the *water table,* the level where the soil becomes saturated.

Water collected like this, deep down below the surface of the earth, is called *groundwater;* it often flows underground, reappearing here and there in the form of springs, and joins the surface run-off flowing into rivers and streams. The much-travelled liquid may be held up in lakes for a time before completing its journey, but it finally flows back into the ocean, ready for the whole process to begin again.

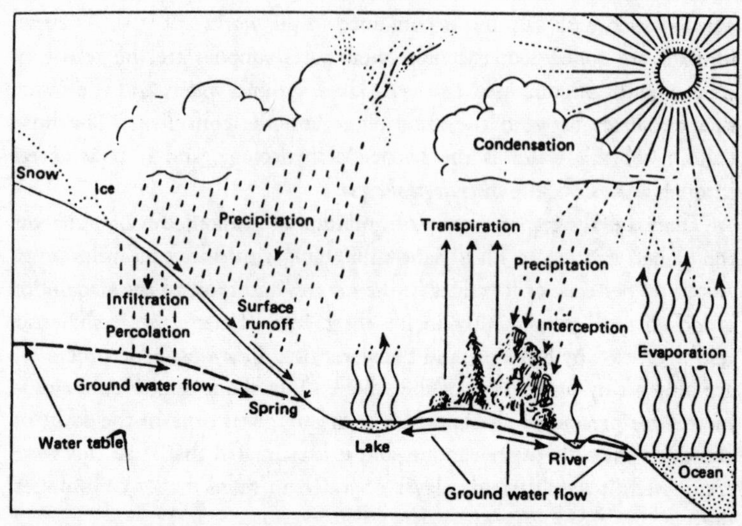

The Hydrological Cycle – the circulation of water through land, air and sea. (From *Hydrology in Practice* by E.M.Shaw)

The Rules of Rain

'It never rains but it pours', according to the old saying. Be that as it may for *hoi polloi*, readers of this volume will be aware that meteorologists are fussy folk, and would never allow a woolly word like 'pours' to mar their ever-vigilant vocabulary. It never 'pours' on weatherpeople – but the rain that wets them may be 'heavy', 'moderate' or 'light'.

They have, of course, reduced it all to numbers, and defined each rate in bands of so-many millimetres per hour. But, broadly speaking, *moderate* rain falls fast enough to form puddles rapidly; *light* rain does not; and *heavy* rain is a downpour which makes a roaring noise on roofs, and makes a fine spray as each drop splashes on the wet surface of a concrete road.

All this rain, whatever its taxonomy, falls from clouds. Clouds consist of tiny particles of water – or, if it is cold enough, of ice – and these particles are very small indeed. A typical cloud droplet of liquid water is only 20 micrometers – or one fiftieth of a millimetre – in diameter, and there are something like one thousand million of them in a cubic metre of the average cloud. They are so small and light that the air around them offers great resistance to their movement; they tend to float in the atmosphere rather than to fall earthwards, which is just as well, since if they did not, each newly formed cloud would collapse quickly onto the ground below!

But sometimes clouds manage to produce bigger and heavier drops which enjoy no such immunity from the law of gravity. This comes about as the tiny droplets of water collide with each other, and merge together to form larger ones. As the drops get bigger, they drift downwards, and since those of different sizes fall at different speeds, the tendency for collision increases, and the result is larger and larger drops. If the cloud is thick enough to allow the drops to become sufficiently big, they eventually fall to the ground as rain.

Raindrops rarely exceed about 5 mm. in diameter. This is because the larger drops become deformed as they fall through the air, and disintegrate into a number of smaller ones, which puts an upper limit on drop size. As a rough rule of thumb, when a raindrop hits smooth cement – such as a footpath – and spreads across it, the spot it leaves has a diameter about seven times larger than that of the original raindrop.

Counting the Drops

The first chapter of *Ecclesiasticus*, as many readers will be aware, deals with the divine origins of all wisdom, and advises strongly against over-estimating one's own cerebral capabilities. 'Who can count the sands of the sea?' demands the author, 'the drops of rain that fall, or the days of all eternity?' Let us throw down the gauntlet! The first and third of these conundrums are outside the scope of this volume, but it is possible to suggest a simple methodology by which the second question might be answered. Let us phrase the problem thus: how many raindrops fall in your back garden during a light shower lasting approximately thirty minutes?

To find the answer we must make a few assumptions. Let us assume, for example, that your garden is 30 ft long and 30 ft wide. This, if you work it out, means that it has an area of about 84 million square millimetres.

Now we know from climatology that in a light shower lasting half an hour, typically about 0.5 mm. of rain hits the ground. This means that if the water stayed where it was, without evaporating, soaking into the soil, or disappearing down a drain, the garden would be covered to a depth of half a millimetre, assuming, of course, that your garden is perfectly level, which I am sure it is. Simple multiplication now tells us that a volume of approximately 42 million cubic millimetres of rain accumulates during the thirty minutes.

Meteorologists have another interesting fact at their disposal: they know the size of the average raindrop. Droplets of very light drizzle – so small that when they fall on a liquid surface they do not cause a splash – have a diameter of about half a millimetre. At the other extreme, the largest raindrops of a very heavy thunder-shower have a diameter somewhere between 4 and 6 mm. But let's assume the norm; let us assume that our 30-minute shower consists of raindrops of about 2 mm. in diameter – a fairly average figure. This means that each raindrop has a volume of a little over 4 cubic millimetres.

And now for the dénouement. If we divide the volume of the average raindrop into the total volume of water which falls on the garden during the 30-minute shower, the answer should be a near approximation to the number of raindrops making up the shower. By my reckoning, it seems to come out at a little over 10 million.

A Rising Start

I seem to remember that there was once an effective slogan, intended to encourage greater safety in the work-place, which went something like: 'Accidents don't just happen; they are caused.' The same could be said of showers. This is not to suggest that a careless citizen may unwittingly bring about a sudden downpour. It means that even when conditions are exactly right, it takes a special circumstance to trigger off a shower.

Rain falls from layered cloud; even though the rain may be intermittent in character, the cloud from which it comes is spread out over a wide area, and is usually associated with a front. Showers, on the other hand, are localized phenomena. They fall from individual clouds, each one anything from a few hundred yards to a mile or two across; in between are patches of brightness, or even of blue sky. The shower clouds are the result of vertical currents in the atmosphere, jets of air gushing upwards like a fountain in slow motion. As the air rises it sometimes cools below its condensation point, a cloud forms, and in due course the coalescing drops grow big enough to fall to earth as a shower.

These rising currents of air are maintained whenever the atmosphere has a suitable thermal structure – whenever temperature decreases rapidly with height. But something has to start them in the first place. Two situations are common, one in the winter and the other in the summer.

The first is frequent in wintertime, when cold air is carried towards us from the direction of Iceland or Greenland. As it flows down from the north-west to lower latitudes, it travels over seas which are progressively warmer. The water heats the atmosphere in contact with it, making it lighter and more buoyant, so that volumes of air in the lower layers tend to float upwards, triggering the vertical currents. Winter-time, therefore, often brings frequent showers near coasts with on-shore winds, but few elsewhere.

In summertime the trigger for showers is usually supplied by the heating of the land – and hence the air immediately above it – by the hot sun during the afternoon. Since this effect is more pronounced inland than near the coast, summer showers are often an inland phenomenon.

Words for the Would-Be Weatherwise

The Oxford English Dictionary tells us, with quite remarkable precision, that the word 'forecast' has been in use with the special meaning of weather prediction since 1673. But weatherpeople prefer a different story. In meteorological circles it is generally assumed that the special meaning of the word was invented in 1860 by Vice-Admiral Robert FitzRoy who had been given the special task of providing 'storm warnings' for naval and merchant ships around the coasts of these islands, and was anxious to avoid 'the somewhat unfortunate connotations attaching to such terms as *prognostic* and *prophecy.*'

Whether true or not, the story suggests that from the very earliest days of the science, meteorologists have gone to great lengths to find exactly the right word to suit the occasion. After all, even a totally accurate weather forecast loses most of its value if it is expressed in obscure or somewhat ambiguous terms. Let's look at a few of the words which are used regularly with 'rain' or 'showers', to describe how extensive they are or how frequently they are likely to occur.

'Isolated' means very few, as in 'isolated showers', or 'isolated patches of rain'. 'Scattered', on the other hand, implies rather more activity. In both cases the rain or showers can be expected to occur here and there over the specified area, but not everywhere. Those who like labels would say that 'scattered' describes the *spatial distribution*; 'occasional' describes the *temporal distribution* of precipitation. Occasional showers can be expected now and then in all places, but the expression also implies that they will fall for less than half the time.

Rain is usually accompanied by cloudy conditions, which continue even when there are temporary breaks in the rain; showers, on the other hand, are separated from one another by bright or sunny intervals. 'Bright' is used when patches of blue sky are likely, but cloud is expected to cover the sun for most of the time. 'Sunny', on the other hand, implies that there will be extended periods when the sun will be visible.

No great international authority, scientific or otherwise, has decreed what shall be understood by all these terms. They are intended, by and large, to be interpreted – albeit fairly precisely – in the same way as they would be in everyday speech, or as they may have been defined in a dictionary, perhaps as long ago as 1673!

Water, Water Everywhere ...

We have no idea when the first rain-gauge was set up. We do not know who had the bright idea, or even in what country the whole rigmarole began. We are told that the Greeks kept systematic weather records of a kind as early as the fifth century BC, and also that quantitative rainfall measurements were made in Palestine in the first century AD. Be that as it may, the design of the rain-gauge has varied little since the earliest times and today they are deployed in great numbers in virtually every country in the world.

The standard rain-gauge has a funnel five inches in diameter to catch the water. The catch drains into a bottle, which is then assumed to contain the amount of rain which has fallen on the 5-inch circle above. Elementary mathematics yields the equivalent depth of water.

Several factors diminish the accuracy of the instrument. Firstly, some of the rain adheres to the sides of the funnel and may evaporate without entering the bottle at all. Then, if the rain is heavy, some of it is lost by splashing out of the funnel, or rain which falls on the ground nearby may splash in. And by far the greatest error arises from the wind; even the rain-gauge itself, projecting above the ground, causes swirls and eddies in the local airflow which result in a loss of catch.

Special care is taken in the siting and the installation of the instrument to minimize these difficulties. Even so, meteorologists are aware that rainfall readings are often inexact by 5 to 10 per cent. But the statistics obtained give a good idea of the variation of rainfall from place to place, and an approximation to the real rainfall which is acceptable for most purposes.

A Dublin Deluge

In the six-hundredth year of the life of Noah, in the second month and on the seventeenth day of the month, all the fountains of the great deep were opened up; and the flood-gates of the heavens overflowed exceedingly and filled all the face of the earth; and the waters prevailed beyond measure.

So says the seventh chapter of *Genesis*, and the description could aptly be applied to events in Dublin on 11 June 1963. It was the occasion of the most dramatic thunderstorm ever to affect our capital city.

The storm of June 1963 was remarkable for the great amount of rain which poured down on Dublin over a relatively short period. The south-side of the city was the worst affected; during the 24-hour period ending 11 p.m. on June 11th, a rainfall station at Mount Merrion recorded 184 mm. of rain, over 80 mm. of this falling in the single hour between 2 and 3 p.m. The storm was local to Dublin, and indeed somewhat less dramatic in other parts of the city, although substantial amounts of rain fell all over the capital during the day.

Let us try to put the cloudburst in perspective. In an average June most parts of Dublin experience a total of something between 50 and 60 mm. of rain over the entire month, significantly less than the 80 mm. which fell in that single hour in 1963. Moreover, using rainfall data collected over the years, it is possible for meteorologists to use statistical methods to calculate for a particular place the 'return period' of rainfall of a given intensity. Such calculations suggest, for example, that a fall of 15 mm. in a single hour could be expected to occur in Dublin about once every 5 years; 20 mm. per hour might be expected every 10 years or so; and 35 mm. of rain in a single hour should only happen once every 50 years. One might almost say that a downpour of 80 mm. in an hour should never happen at all – but it did!

It is possible, of course that heavier rain has occurred in Ireland from time to time, but passed into oblivion undocumented. Quantitative assessment of the amount of rain which falls in any heavy cloudburst depends upon on a rain-gauge being in operation at the right place and at the right time to provide us with precise figures. Many spectacular falls over the years may have escaped detailed analysis by climatologists, their torrential nature evident only to those unfortunate enough to have been caught underneath them, or to have become victims of the flooding which frequently ensues.

Troublesome Waters

In a climate such as Ireland's, a great deal of technological effort is expended by the civil-engineering fraternity on the fundamental problem of keeping out the rain. Heavy rain alone is troublesome, but when combined with the penetrating force of a strong wind – a phenomenon known to meteorologists as *driving rain* – the problem is magnified many times. Faced with the 'to-and-fro-conflicting wind and rain', detailed research pays dividends in finding the materials best able to 'bide the pelting of the pitiless storm'.

In calm conditions, raindrops fall vertically at a speed which depends upon their size. When the wind blows, however, the drops are also carried horizontally; with very strong winds, the rain is swept along at angles close to 45 degrees, and when this happens a vertical wall facing the wind receives a substantial wetting.

The effects can be significant. During a prolonged spell of wind-driven rain, water can quickly penetrate a wall through tiny cracks; the structural fabric of the building may be affected, and damage may be caused to the interior decoration. At the very least, the absorbed rainwater increases the thermal conductivity of the walls, and by lowering the temperature inside may lead to condensation on the inner face. It also increases the fuel consumption required to keep the house warm.

Difficulties of this kind are not insurmountable, and steps can be taken to minimize them in places where driving rain might be a problem. Careful choice of site and orientation, for example, reduces the exposure; special architectural features – such as large roof over-hangs – can provide good protection; and special methods of construction, or a protective coating applied to the finished surface, help to avoid unpleasant consequences.

But it is first necessary to know the extent of the problem. With this in mind, meteorologists have devised what they call the *driving rain index* – the annual average wind speed multiplied by the annual average rainfall at a particular spot – and this quantifies the susceptibility of an area to driving rain. As one might expect, the driving rain index for western counties of Ireland is high compared to elsewhere in the country, but there are strong variations with the local topography, mountain areas experiencing much higher values than sheltered spots in the vicinity.

Thoughts on Snow

Most raindrops have a lost youth, a hidden, unsuspected but eventful history which is by no means obvious as they fall to the ground in their old age as troublesome drops of water. Each is born out of a tiny ice crystal, which in due course becomes part of a complex snowflake. Usually the snowflakes melt before they reach the ground, but sometimes, if the weather is cold enough, they survive in this form, and the result is a fall of snow.

Even the largest ice crystals are only a millimetre or two in diameter; they come in a variety of shapes, like needles, prisms stars and plates, most of them hexagonal. Given the right conditions, these crystals *flocculate*; they aggregate together to form the complex patterns of the familiar snowflake. The largest flakes occur when the temperature is just slightly above zero, since the crystals are then slightly wet, and adhere together more easily; at very low temperatures, there is less aggregation, and so the individual snowflakes are smaller.

For the same reason – because its temperature is rarely far from zero – Irish snow is very good for snowballs. To form a snowball, it is necessary for individual flakes of snow to cling together, and this happens most readily near the freezing point. As every schoolboy knows, it is the pressure applied to the snow by the hands which causes it to coalesce into a nice firm ball.

The melting point of ice is normally taken to be 0°C, but this is strictly true only at atmospheric pressure. If pressure is significantly increased, the melting point is lowered. In the process of making a snowball, compression causes quite dramatic increases in pressure near the pointed ends of the ice crystals in the snow, and this in turn causes localized melting to take place at the temperature of the snow, perhaps one or two degrees below zero. When the pressure is removed, the water produced by the melting freezes again, and rigidly joins neighbouring crystals. This process is also at work in the formation of a crust of snow on the soles of our shoes when we walk on fresh snow.

Freshly fallen snow is a very good thermal insulator, because of the large volume of air contained between the individual crystals and flakes. This explains how animals can survive for such a long time, even when buried deep in snow-drifts at temperatures which would normally be fatal. Small plants are similarly protected by snow from the worst ravages of the harsh frosts of winter.

Old Winter's Tales

One winter's morning in 1786 the gentry of our metropolis awoke to find the leader writer of *Faulkner's Dublin Journal* to be in fighting fettle. 'The pernicious custom of throwing snowballs has arrived at an intolerable height,' he thundered; 'no less than a dozen decent persons have been desperately wounded by stones and brickbats wrapped up in these missile weapons of barbarous amusement.'

And in the following year, the *Dublin Chronicle* was more provocatively vociferous on the same subject:

A gentleman passing through Marybone Lane was hit by a fellow in the face with a large snowball, upon which he immediately pulled out a pistol, pursued the man, and shot him dead. Those deluded persons are therefore cautioned against such practices, as in similar circumstances they are liable, by Act of Parliament, to be shot, without any prosecution or damage accruing to the person who should fire.

Nowadays, the legal and social climate is more benign towards would-be snowballers, but then the opportunity for them to practice their mischievous art occurs a good deal less frequently than it used to. Snow was a much more common occurrence in this country some two or three hundred years ago during what has come to be called the 'Little Ice-Age'.

In mediaeval times Northern Europe enjoyed a comparatively obliging climate, somewhat warmer than we know today. But a sudden change took place in the middle of the sixteenth century. Temperatures dropped dramatically, and the period from about 1600 to 1850 was the coldest since the ending of the last ice-age, 10,000 years before. During this period the average global temperature was between 1 and 1.5 degrees below today's norm, and in the coldest years – notably 1695, 1725, and 1740 – the average was up to 2 degrees lower than present-day values.

In Britain and Ireland, the winters were long and very severe, and the summers cold and wet. The consequences, of course, were not all bad. During the long winters, ice-skating was much in vogue, and the populace could disport themselves upon the frozen rivers in a way which is impossible today. Such scenes are now found only on the traditional Christmas card, or illustrating the novels of Charles Dickens, whose childhood memories would have been firmly rooted in the closing years of that Little Ice-Age.

Washing Weather

The name of Hamilton Smith of Pittsburgh, Pennsylvania, USA, would no doubt have been consigned to cold oblivion were it not for his preoccupation with domestic chores. In 1858 Smith built what is believed to have been the very first mechanical washing machine. It was a crude device, operated by turning a crank at the side of the apparatus which rotated paddles inside the tub, and perhaps it is a measure of its effectiveness that it was another fifty years before such labour-saving devices began to be taken seriously.

Washing machines come to mind because they provide the raw material for what proves to be a very difficult exercise in changeable weather conditions: the drying of clothes. In scientific terms, the problem is to change water from the liquid into the vapour state through the process of evaporation. It is an energy-intensive process; by way of illustration, *five* times as much heat is required to vaporize a pint of water, as is needed to raise the temperature of the same pint of water from zero degrees Celsius to boiling point. Evaporation therefore increases when abundant energy reaches us from the sun: when there is bright *sunshine.*

The *temperature* of both the air and the evaporating surface is also important. The higher the temperature of the air, the more moisture it absorbs from the wet clothes; if the temperature of the evaporating water is high, it vaporizes more easily. Another major factor is the humidity of the surrounding air. Dry air has plenty of spare capacity for moisture; very damp air can only accommodate a little extra water before becoming saturated, at which point it refuses to collect any more.

When water evaporates in calm conditions, the still air closely adjacent to the evaporating surface gradually becomes more humid, until finally it is saturated. If however the air is moving, the rate of evaporation is increased by the constant arrival of a fresh supply of drier air. *Wind* improves evaporation, and drying is better in exposed areas than in sheltered places where the air tends to stagnate.

Assuming, therefore, that you have a choice in the matter, best drying can be achieved by erecting your clothes-line in a well exposed position and hanging out the clothes on a warm, sunny, windy day when the humidity is low. But then, perhaps, you do not need the help of a meteorologist to work all that out anyway!

Dew Process

'When the sun sets the air doth drizzle dew', declared Old Capulet. It was his strange Shakespearean way of explaining the tears of his daughter Juliet, weeping, he assumed, over the untimely death of cousin Tybalt. As it happened, Capulet missed the point; Juliet's thoughts were on her lover, Romeo Montague. Life and love are never simple, like meteorology.

Sometimes even the weather can mislead. Dew drops on a sunny morning do indeed appear as if they must have dropped from heaven. But with the sky clear and cloudless, where do they come from, these little jewels of the dawn?

Like drizzle or rain – or indeed a cloud, or fog – dew forms when the air contains more moisture than it can comfortably hold. But in the case of the first four phenomena, moisture has condensed into little drops of water suspended in the air; dew, on the other hand, is deposited *directly* onto a cold surface. It does not fall from anywhere; it forms *in situ*.

On a clear night with cloudless skies, heat from the earth is lost to space by long-wave radiation, causing the ground temperature to fall rapidly. As the ground cools, so too does the air in immediate contact with it, and frequently the temperature in this very thin layer falls sufficiently for condensation to take place, and dew forms. Dew occurs only if the wind is very light indeed; if there is even a slight breeze, the resulting turbulence causes the cold air nearest the ground to be mixed with warmer and dryer air above, and the resulting temperature will more than likely be above the point where condensation might take place.

Dew has a preference for certain types of surface. It tends, for example, to form more readily on grass than over bare soil. This is because the earth is a good conductor of heat, so as the temperature of the top-most layer falls, there is a flow of heat upwards from the layers of soil underneath; this makes the reduction in temperature at the surface less than it would be otherwise. The upper leaves of a clump of grass have no such reservoir to draw upon; they become colder, and the dew forms quickly. And for much the same reasons, dew tends to form more readily on good thermal *insulators* – like wood or glass – than on good *conductors* of heat, like metal; in the latter case, the energy radiating from the surface is replaced to some extent by a flow of heat from the interior.

The Details of Obscurity

'On a clear day', according to the song, 'you can see forever!' But of course you can't; rays of light coming towards you through the atmosphere are undergoing a continuous process of attenuation, the extent of which depends on the composition of the air at any particular time. Even if the air were perfectly clear, it is reckoned that maximum visibility would be a mere 150 miles; in practice, to be able to see clearly farther than, say, 40 miles is quite unusual.

Apart from all the molecules of air, which are themselves capable of scattering light of certain wavelengths, the atmosphere contains impurities, suspended particles so small that they settle to the ground only at extremely low wind speeds. If these particles are of *dry* matter – smoke or dust – any obscuration they may produce is known as *haze*. If they consist predominantly of *water droplets*, they result in mist, and if the mist thickens to the extent that visibility falls below 1000 metres, meteorologists define it as a fog.

The presence or otherwise of these small particles in sufficient numbers to reduce the horizontal visibility significantly is largely determined by the amount of vertical motion in the atmosphere at any particular time. If there is little vertical motion, they remain in the lower layers near the ground, and so visibility is reduced. If there are plenty of rising currents of air, the particles are dispersed throughout a very deep layer of the atmosphere, and visibility is good. The weather-map can often tell us what to expect in this regard.

Depressions, by and large, are regions where there is a great deal of upward movement of the air. In the warm sector – the zone between the cold and warm fronts – visibility is often reduced by drizzle or rain, but in the north-westerly airflow behind the cold front, the air is usually very clear indeed. There is often a dramatic improvement in visibility after a cold front has passed; the rising air-currents in the zone behind the front may produce showers, but as a rule the air is pleasantly clear and pure.

Anticyclones, on the other hand, despite the fine settled weather normally associated with them, are regions where there is little movement of the air in the vertical. For this reason visibility is often disappointingly poor, and far-away hills are often obscured by a thick, blue, smoky haze.

Focusing on Fog

Fog has a sinister reputation. Mediaeval fogs were feared as the embodiment of an unhealthy dampness, as catalysts for rheumatic aches and pains, and as evil vectors for every kind of ague and fever. Even today, in cities where unchecked pollution leads to smog, mortality statistics lend some credence to these ancient fears.

Fog, as previously defined, is composed of tiny droplets of water suspended in the air. The droplets are so small that it would take seven thousand million of them to make a single teaspoonful of water. In clean white 'country' fog each droplet is pure water, with just the merest trace of salt, the salt often being the nucleus around which the drop has formed in the first place. Yellowy 'city' fog, on the other hand, is often found to have formed around tiny particles of soot.

When an area of high pressure lies in the vicinity, it provides the light winds, clear skies and stagnant atmosphere necessary for fog formation. The main requirement is that moist air should be cooled to what is called its *dew point*, the temperature at which it is no longer capable of accommodating all the water vapour it originally contained. In such circumstances some of its moisture is expelled in the shape of droplets of water.

Clear skies are favourable because it is in such conditions that ground temperature at night drops most quickly, as the unprotected earth loses heat rapidly out to space; a layer of cloud overhead acts like a blanket to retain the warmth. Calm conditions are important because wind stirs up the atmosphere, and consequently the air cannot remain long enough in contact with the cold ground for its temperature to be reduced to the dew point; it is mixed with warmer air from the layers above.

Anticyclonic fog is by nature an inland phenomenon, because nocturnal cooling takes place only over land. Sea surfaces are much less affected in this way, varying little in temperature from one time of the day to another, although fog, once formed, may then drift out over coastal waters. Fog is often first observed in low-lying places, where the cooler, heavier air has drained downward into hollows in much the same way as water runs downhill. River valleys are favoured spots; in addition to the pool of cold air there is a ready supply of moisture from the river water to facilitate the early saturation of the air above.

A Deep Obscurity

Kipling, like all of Gaul, may be divided into three parts: his time as a journalist in India, which provided much of the material for his prolific output, followed by a period spent travelling and living in the United States, after which he settled down in the south of England. It was there that he would have become familiar with the phenomenon of sea fog, so evocatively portrayed by him in his poem 'Sussex':

> And here the sea-fogs lap and cling
> And here, each warning each,
> The sheep-bells and the ship-bells ring
> Along the hidden beach.

Meteorologists, of course, like dissenting Greeks, have a different word for sea fog; for reasons which I will try to make clear, they call it *advection* fog. But call it what you will, it is a familiar sight along our southern and eastern coasts in spring and early summer.

Advection fog is quite different in character from the *radiation* fog to be found filling the inland valleys of Ireland early on a bright sunny morning. It occurs when warm moist air flows over a cold surface; it can develop only if the temperature of the surface underneath is below the *dew point* of the air, the temperature at which condensation takes place.

Advection fog often occurs over Irish waters in the early part of the year when the sea is still relatively cold. It can develop for two reasons: a breeze may carry warm air from the land out over the relatively cold water; or warm 'tropical maritime' air may approach Ireland from the south, and be cooled by contact with the relatively cold seas around our coast. Either way the temperature of the air is reduced to a temperature at which condensation takes place and the result is fog.

In contrast to the radiation fog which forms over land, sea fog can occur at any time of the day or night, and is not restricted to conditions of light winds and clear skies. Indeed it is more common when there is a noticeable breeze, because its very existence depends on some movement of the air; the air must be *advected* over a cold surface. And it is very much a coastal phenomenon; summertime sea fog usually dissipates quickly on moving inland, because the warm ground underneath heats the air above its dew point, and allows it to re-absorb the moisture.

Kept in Suspense

Above your head at this very moment, enormous amounts of water hang suspended in the sky in the form of clouds. A small cumulus – those little tufts of cotton wool so common on a fine day – may hold anything from a hundred to a thousand tonnes of water. The great towering cauliflowers which produce heavy showers and thunderstorms – cumulonimbus clouds – are veritable mountains of water and ice; if the whole body of water contained in such a cloud could be put into a gigantic bucket and weighed, it would turn the scales at something like a hundred thousand tonnes. And the amount of water in a typical mid-Atlantic depression is so vast that no meaningful figure for it could be calculated.

These extraordinary statistics prompt a very obvious question: how do these great quantities of water manage to stay put? Water, after all, is about 800 times heavier than air, so a Wordsworthian cloud that 'floats on high o'er vales and hills', ought to be, one might think, a physical impossibility. The whole celestial ensemble ought to collapse onto the ground below in a soggy pool.

This, as we know, does not happen. A cloud consists of very fine droplets of water, whose diameters range from a ten-thousandth to a hundredth of a centimetre. Each drop, looked at individually, does in fact fall through the air, but their rate of descent is so small as to be almost imperceptible.

According to Newton's Second Law of Motion, a falling object – left to its own devices – should fall faster and faster as the seconds tick by. But there is another force to be reckoned with; the same object, falling through the atmosphere, is also subject to resistance from the air, and the greater its rate of descent, the greater this resistance. Sooner or later, the point is reached where the tendency for the rate of fall to increase is balanced by the action of the air slowing the object down; it reaches what is called its *terminal velocity*, a constant rate of descent.

Light objects tend to have a rather low terminal velocity, as you may have noticed if you allow a feather to fall to earth. Cloud droplets have a very low terminal velocity indeed; of the order of only a few centimetres per second. This rate of movement is negligible compared to the other motions in the atmosphere – the swirls and eddies of the moving body of air – so unless the drops grow big enough and heavy enough to fall as rain, the cloud stays suspended indefinitely above our heads.

The Advance Guard

The very highest clouds are called *cirrus* clouds, the familiar 'mare's tails' that in fine weather often stand out in bright contrast to the brilliant blue of an otherwise cloudless sky. Cirrus looks like swirls or 'hooks' of cotton wool, fibrous in structure and too thin to cast a distinct shadow or to blur the outline of the sun or moon. It nearly always consists entirely of ice particles, because at the great heights at which it forms – six or seven miles above the ground in mid-latitudes, and half as much again near the equator – the temperature may be –40°C or less.

The wispy appearance of cirrus has a number of contributory causes. In the first place, at those heights the air is so cold that it cannot hold sufficient water vapour to form thicker clouds. Also, when the water vapour is condensed, it is distributed in the vertical over a deep layer of the atmosphere; the wind at the bottom of this layer may not be blowing in the same direction as the wind at the top, a circumstance which gives the cloud its characteristic 'twist' or 'hook'. The tufty appearance is accentuated by the oblique angle at which we are normally obliged to view the cloud.

When cirrus slowly disappears, it is often an indication of fair weather in the immediate future. But when it grows denser, it is frequently the harbinger of rain. The thickening cirrus is a kind of scout or *avant-garde* leading the way for the serried ranks of lower, heavier clouds advancing in its wake, the first visible precursor of an invading front.

Alto-Confusion

The word *cirrus* means a 'lock of hair' or a 'curl', a term which is both imaginative and descriptive. In the case of medium-level clouds, on the other hand – those between about 6000 and 18,000 ft above the ground – the names are prefixed with *alto*, meaning 'high'.

There are two types – or *genera* – of 'alto-cloud': *altostratus* and *altocumulus*. Altostratus is, almost literally, a gloomy wet blanket which nearly always leads to rain. It is dark grey and very uniform, and normally advances steadily like a thickening curtain being drawn across the sky. At first it may be thin, allowing a faint outline of the sun to shimmer through it, ghostly in aspect, as if it were being observed through a sheet of ground glass. It is usually associated with a front, and as the frontal rain comes nearer and nearer, the altostratus grows thicker, gradually obscuring the sun completely in a dark, smooth, threatening layer.

Altocumulus, on the other hand, is brighter and may not necessarily promise rain. A whitish-grey colour, it is distinguished only by its height from another common cloud-form, *stratocumulus*, which occurs lower down in the sky between 2000 and 6000 feet. In contrast to altostratus, altocumulus is non-uniform in appearance; it displays light and dark shadows, caused by the fact that the cloud is not a flat sheet or layer but is composed of 'rolls' or undulations, often arranged in long straight columns across the sky. It is typical of the cloud to be seen on a dry, rather cloudy day, when there is little sunshine but no particular threat of rain.

Both altocumulus and altostratus are composed of water droplets, unlike their more lofty cirriform cousins whose constituent particles are tiny ice crystals. When the sun is shining weakly through a thin layer of altocumulus or altostratus, the light is often deflected by the water droplets in such a way as to form a *corona*, a bluish-white circle of light which, when looked at more closely, can be seen to embody faintly all the colours of the rainbow, from violet and blue on the inside through green and yellow to a pronounced reddish-brown near its circumference.

Medium-sized cumulus clouds, typical of a summer afternoon. (A.J. Aalders, Netherlands)

Massive Cumulonimbus cloud. (A.J. Aalders, Netherlands)

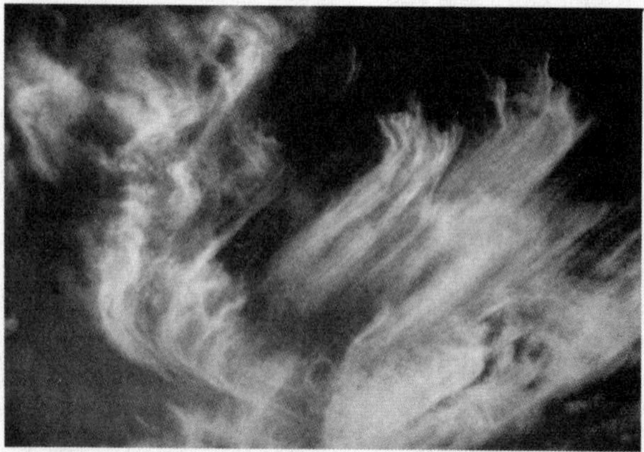

Hooks of cirrus signal an advancing front. (Royal meteorological Society, Clarke Collection)

A broken layer of altocumulus covers the sun. (A.J. Aalders, Netherlands)

Stratus cloud shrouding a mountain-top on a misty day. (R.K. Pilsbury)

With the approach of rain, a weak sun gleams through a layer of altostratus with small broken cumulus below it. (A.J. Aalders, Netherlands)

The Spectacle of Clouds

Altocumulus lenticularis is not a disease, nor indeed is it a newly discovered plant. It is the classical name given to a very pretty type of cloud which is sometimes to be seen in the vicinity of mountains or high hills.

Lenticular cloud, as it is more commonly known, has sharply defined edges and thin, somewhat pointed, ends. The total effect resembles that of an elongated almond, or as its name implies, the cloud has a cross-section rather similar to that of an optical lens. It is often arranged in well-separated bands stretched across the sky parallel to a range of mountains. The most striking thing, if you observe the cloud closely, is that it does not move with the wind; it remains stationary, often for hours on end, locked in position relative to the ground below.

It is the mountains themselves which are responsible for the cloud's formation. As the large obstacle disturbs the smooth flow of air, wave-like undulations in the airflow form downstream. The resulting waves are called *standing* waves, because even though the air in which they form is often moving quite rapidly, the position of each crest and each trough remains fixed in position relative to the mountain.

Lenticular cloud forms in these conditions if the air is moist. Air cools as it rises, and if it contains sufficient moisture the fall in temperature may be enough to cause its relative humidity to reach 100 per cent, in which case condensation occurs. It sometimes happens that condensation takes place in this way as the air rises to the crest of each standing wave but as it descends again into a trough, the air becomes warmer and capable of re-absorbing the droplets of water moving along with it. The result is a series of long shallow cloud-bands arranged at right angles to the flow of wind, each one marking the position of the crest of one of a succession of standing waves.

In fact, when you look at a lenticular cloud, it is not the *same* cloud you see all the time; the cloud is constantly being formed on the windward side of each wave, as moisture condenses from the rising air, and disappearing on the lee side, as the air descends and the water droplets evaporate again. A nineteenth-century meteorologist called J.F. Daniels described it rather nicely: 'The apparent permanency and stationary aspect of a cloud is often an optical deception, arising from the solution of vapour on the one side of a given point, while it is precipitated on the other.'

A Child of the Sun

When the weather is calm and warm, and the sky relatively clear, surface heating by sunshine produces ascending columns of air, here, there and yonder around the countryside. If the atmosphere is relatively humid, a cloud forms near the top of each of these invisible pillars; a bulbous, rounded cloud with a flat base, resembling a clump of cotton wool or a giant cauliflower.

This is a *cumulus* cloud. It is brilliantly white where the sun shines on it, and blue-grey to dark over the shaded portion, except along the edge nearest the sun where the hem of the fabled silver lining can often just be seen. Literary experts are unanimous in the view that when William Wordsworth reclined on his couch contemplating his 'cloud that floats on high o'er vales and hills', it was fair-weather cumulus he had in mind.

Looking around over the landscape, it can be seen that some of these clouds tower higher than their neighbours. But the bases of all the clouds, big or little, are at more or less the same level; the level at which the rising surface air will have cooled sufficiently for saturation to be reached and condensation to begin. As a cumulus grows in the heat of the day, the air within the cloud is in a state of constant vigorous agitation; its sides and domes change shape continuously, bubbling upwards and outwards with a bright assertive effervescence.

Cumulus is a child of the sun, a product of the fine weather. It does not occur when the sky is overcast, nor at any time of considerable wind, because wind churns up the atmosphere and prevents the establishment of the engendering columns of ascending air. At times, because of the abundance of moisture available, it forms a chain of balloons, as it were, to mark the course of a large river. Over land it is more numerous by day than at night, and over the sea it frequents islands far more than it does the open ocean.

The stronger the heat of the sun, the greater the convective activity, and the larger and more luxuriant the huge cauliflower tops of the growing cumulus. But towards sunset the convection stops, and the cloud, no longer having a supply of moist rising air to support it, begins to sink. The sinking air is warmed by compression, and since it can now support more moisture, the cloud starts to evaporate. Ultimately, in the gathering twilight, it disappears completely or flattens out into a dull layer of grey-pink stratocumulus.

Nightlights

The highest clouds normally visible are seldom more than five miles above the ground. But *noctilucent* clouds are an exception; they occur in a narrow zone about fifty miles up, at the coldest level of the atmosphere called the *mesopause.*

These are very rare clouds. Their texture is such that they are not seen during day, because the sun just shines right through them. They are so thin that they are visible only at twilight, when the sun is between 6 and 16 degrees below the horizon. In such circumstances, the clouds are still sunlit because of their great height, while the atmosphere below them – and any ground-based observers – are shrouded in darkness.

Noctilucent clouds generally appear low in the northern sky, and have a silvery-white appearance, often with a bluish tinge. They can be distinguished from ordinary high clouds by the fact that they are not tinted by the usual red glow of twilight; they also stand out brightly against the afterglow, in contrast to the familiar high cirrus clouds which appear dark. Noctilucent clouds are a summer phenomenon, seen in the northern hemisphere mainly in the months of July and August, and most frequently between the latitudes of 50 and 65 degrees north. This, in our part of the world, means roughly from Ireland to Iceland.

These ephemeral ornaments of the evening sky were noticed for the first time near the town of Kissingen on 8 June 1885 by a German meteorologist called T.W. Backhouse. It is something of a mystery that there were no reported sightings prior to then. Their discovery may have been due to the fact that the sky was being closely observed around that time, because of spectacular sunsets associated with the recent eruption of Krakatoa in the East Indies. On the other hand, it may well be that there were simply none to be seen!

Their formation depends on critical levels of water vapour at high levels in the atmosphere. Water vapour at these heights is known to be a by-product of the breakdown of *methane,* and levels of methane in the atmosphere have been increasing steadily for 150 years or more. The increasing abundance of this gas may well be the reason why sightings of noctilucent clouds appear to be becoming less rare nowadays.

The Hail Story

Although they may occur independently, hail and thunderstorms go together. They are produced by the same type of cloud, the large towering *cumulonimbus* clouds formed by powerful up-drafts in the atmosphere. In winter-time their origin can usually be traced to cold air from the vicinity of Iceland which has swept down from the north-west, and is heated near the ground by the relatively warm sea and land underneath. In summer, it is the strong heat of the sun which heats the ground itself and provides the stimulus. In any event, this warming near the surface make the lower layers buoyant, and causes the atmosphere to bubble upwards like boiling water in a saucepan.

A hailstone begins its life as a small particle of ice in the upper reaches of a cumulonimbus. If the cloud is a small one, the hailstone may simply fall straight to earth, perhaps even melting on the way and falling as a raindrop. But if the cloud is a big one, the hail may be carried up to the top again by an updraft. As it moves within the cloud, the hailstone encounters tiny drops of 'supercooled' water: water below the freezing point, but not yet turned to ice. Such droplets freeze on impact with the hailstone, and make it larger.

A hailstone may make several trips up and down the cloud before falling to earth and on each journey it accumulates an extra layer of ice, often opaque in appearance because of the little bubbles of air trapped in between. Very large hailstones, if sliced in half, can be seen to be composed of a great many layers of ice – rather like an onion – reflecting the mechanics of their formation.

Hailstones above a centimetre or two in diameter are rare in Ireland but common in other parts of the world. In the United States they often grow to three or four inches, and cause serious damage to crops. And as an extreme example, giant hailstones weighing 2.25 lbs are reported to have killed ninety-two people in the Gopalganj district of Bangladesh on 14 April 1986.

Raining Cats and Dogs and…

The treatise 'De Pluvia Piscium' (On the Rain of Fishes), written by Athanasius over 1500 years ago, contains the first known account of that strange and rare phenomenon: a shower of small animals descending from the sky in rain. Samuel Pepys knew about it too; in his diary entry for 23 May 1661 he records that 'Elias Ashmole, the antiquary, did assure me that frogs and many insects do often fall from the sky, ready formed.'

But there are reports closer to the present day. One memorable and well-documented fall of frogs occurred at Selby in Yorkshire in 1844 when people were able to hold out their hats to catch the falling reptiles. In 1877 there was reputedly a shower of snakes in Memphis, Tennessee, thousands and thousands of them ranging from a foot to eighteen inches long. On 17 June 1939 the superintendent of the local swimming pool in Trowbridge in Wiltshire was running for shelter from a sudden downpour when 'I turned and was amazed to see hundreds of tiny frogs falling on to the concrete path surrounding the pool.' And as recently as June 1984 the owner of a service station in the north of England, thirty miles from the sea, arose one morning to find winkles and starfish covering the forecourt of his garage and the top of its high canopy; the winkles were salty, and many of them were still alive!

So what do meteorologists make of all these strange events? They are somewhat sceptical of most reports. They prefer to believe, for example, that apparent showers of frogs may have been caused by the reptiles simply having been tempted in large numbers from their hiding places by the sudden downpour of pleasant rain.

For authenticated occurrences though, the most likely explanation is a *waterspout.* Waterspouts are similar in nature to tornadoes, but are a good deal less violent and form over water, usually in thundery conditions. Like a tornado, the whirling vortex of a waterspout is capable of sucking up solid objects in its path; these may then be suspended for some considerable time in the powerful updrafts of a thunderstorm, before being deposited quite a distance away in the course of a heavy thundery shower.

VI *An Optical Assortment*

The Lightning Lottery

It is reputed to have been that old rascal Benjamin Disraeli who distinguished three kinds of lies: 'lies, damned lies, and statistics'. And it is true, of course, that figures can be moulded to suit almost any argument. They can be frightening or reassuring, depending on how you look at them. Take, for example, the chances of being killed by lightning.

The average thunderstorm, we are told, contains ten times as much energy as the atomic bomb which devastated Nagasaki in August 1945. From such a storm come bolts of lightning which can generate temperatures as high as 25,000°C. We are also informed by the statisticians that at any given moment nearly 2000 such thunderstorms are in progress over the earth's surface, and that lightning strikes the ground 100 times every second. How do we survive at all, with the odds so heavily piled against us?

But there are reassuring figures to be found if you look around for them. According to studies, the odds against an individual in England or Wales being killed by lightning in any one year are about 13 million to one. The odds in Scotland are around 17 million to one, and in Ireland the risk is quantified at 9 million to one. Judged purely on these figures, Ireland might seem a slightly precarious place to live. But the statistics, if not exactly lying, must at least be suspected of telling a little fib; it seems likely that the lower odds are explained by our lower population, rather than by any national susceptibility to fatal lightning strikes. And even at their worst, the chances of being killed by lightning are clearly very small indeed.

There is, however, a sexist dimension to the statistics of fatal lightning: fortune, it seems, favours the gentler sex! The odds against a female being killed by lightning in Ireland in any given year are reckoned at 15 million to one, compared to odds in the case of males of a mere 6 million to one against. The real reason, of course, is not a bias on nature's part, but the fact that men, both by occupation and in recreation, tend to spend more time out of doors than women do.

Inspiration from a Flash

Man has always had a wholesome respect for the destructive power of the flash of lightning, and this ancient fear is reflected in the folklore and superstitions of many cultures. Indeed even to this day the words describing thunder and lightning are mildly sacrilegious in some languages. Remember the Biggles-cornered villain, totally out-manoeuvred by our hero, heard to mutter in his desperation – 'Donnerwetter! Donner und Blitzen!'

The North American Indians believed that lightning was produced by the magical thunderbird, its vivid plumage and beating wings as it dived from the clouds providing the lightning flash and the clap of thunder. Tangible evidence of its visitation remained as the mark of its claws on the scarified bark of damaged trees and splintered tepee poles. To the ancient Greeks, however, the lightning flash was an indications of the displeasure of their chief god, Zeus. There was protection from this wrath; it was known that the laurel bush – sacred to Zeus's son Apollo – was immune from lightning strike, and so a laurel wreath was a guarantee of safety in a thunderstorm.

The Roman Emperors wore the laurel wreath for different reasons, but they had their own superstitions about lightning. To them a lightning flash passing from left to right across the sky was a favourable omen for affairs of state; but headed towards the left it was – quite literally in the Latin language – a *sinister* sign. It showed that Jupiter did not approve of current undertakings.

Mediaeval seers gained even greater insight. Indeed by the sixteenth century Leonard Digges in his book *Prognostications of Right Good Effect* was able to give a day by day prediction of the effects of thunder and lightning. Sunday's thunder, according to Digges, brings the death of learned men, while Monday's thunder tells of death of women. Thunder on a Tuesday, by contrast, is a good omen; it signifies an abundance of corn, and a good harvest.

On Wednesday thunder is bad news for those of easy virtue; it bodes 'the deathe of harlottes and other blodshede'. Things look up again on Thursday, a thunderstorm on that day being once again a hint of opulence to come, but should it occur on a Friday it foretells the death of a great man 'and other horrible murders'. But perhaps most ominous of all is thunder on a Saturday; it brings 'a generall pestilent plage and certain deathe'.

Ringing the Changes

Tintinnabulation was widely used as a meteorological technique in mediaeval times. It was believed that ringing the church bells would protect crops from hail, and ward off the evil spirits of the storm, but even in its day, it was a controversial practice.

Hail was a major problem. Many places throughout central Europe, particularly the mountainous areas in the centre of the continent, are subject to very severe and frequent hailstorms. The large lumps of ice strip the leaves from the vines, batter crops and cereals to a useless pulp, and cause irremediable damage to young citrus fruits. It was no uncommon thing for a peasant to see a whole year's livelihood swept away in a few minutes, and anything which might avert such a disaster was well worth a try.

The accepted wisdom of the time was that hailstorms were caused by dark spirits in the air. By frightening away these demons with a loud noise, the worst of the tribulations might be avoided – and what better way to create a din than to ring the church bells as loudly as possible? So orthodox was the practice that Pope Urban VIII authorized a special prayer for use by bishops when consecrating the bells:

> Grant, O Lord, that the sound of this bell may drive away harmful storms, hail, and strong winds, and that the evil spirits that dwell in the air may by Thy Almighty power be struck to the ground.

Nothing, alas, is that simple; human nature frustrated these attempts to harness divine acoustic intervention. Firstly, many landowners some distance from the bells believed that the noise diverted hailstones from the vicinity of the church onto their land. They were understandably annoyed. Another serious problem was the number of bell-ringers who were killed by lightning; hail is often associated with thunderstorms, and church towers were particularly vulnerable to lightning-strikes. And at the other extreme, some villages complained that they were being unfairly treated by the authorities, because they had no bells to ring at all!

In due course Charlemagne was thus obliged to issue an edict forbidding the use of church bells for this purpose. But even then, no one was satisfied; communities which suffered from hail damage complained that they were being deprived of the right to defend themselves. Many defied the edict, even to the extent of overpowering their unfortunate pastors who tried to enforce the Imperial law.

The Inflammation of Castor and Pollux

An rud is annamh is iontach (an event that's rare is always wondrous strange), they used to say. So it was with the ghostly light of St Elmo's fire, a seldom-seen display of atmospheric electricity which, by and large, was welcomed as a friendly omen by mariners in ancient times. The flame-like streamers were three or four inches long, and came in a faint but luminous glow from the masts and rigging of the old sailing ships. St Elmo's fire can still be seen if the times are right; it flows from lightning conductors on top of tall buildings, or from the weatherman's wind vane, or now and then even streams from the loose ends of human hair.

Nobody is quite sure who old St Elmo was, or even if he ever existed. Some believe the name to be a contraction of St Erasmus, a fourth-century bishop whose fort even to this day guards the entrance to the Grand Harbour of Malta. But there was also a thirteenth-century Dominican preacher of that name, who worked among sea-faring folk around the coasts of Spain. Whatever the origins of the name, St Elmo's fire has a simple scientific explanation.

It occurs when the air is thundery and the clouds above are highly charged with static electricity. Sometimes the tension between the earth and clouds is such that an electric current wants to flow from one to the other. Now and then, rather than bursting forth as a flash of lightning, the charge just 'dribbles' upwards from the highest point in the vicinity; it is this electrical perspiration which causes the surrounding air to glow.

For sailors in bygone days St Elmo's fire was an influential portent of things to come. If only *one* stream of light appeared, it was believed to be a manifestation of Helen of Troy, she 'whose beauty summoned Greece to arms, and drew a thousand ships to Tenedos.' Considering all the trouble she caused on that occasion, mariners came to the very reasonable conclusion that Helen's comeback boded ill; a single flame of St Elmo's fire foretold storm and tempest.

But if *two* flames appeared, the signs were good. The twin flames were believed by sailors to be, hardly an incarnation, but what might be called an 'inflammation' of Castor and Pollux, the semi-divine twin brothers of Helen, who tried so hard to rescue her. The two had done their best, had been given a protective role over ships at sea, and their sudden re-appearance – *annamh* as it might be – was seen as a benign prognosis of pleasant calms to come.

They Seek It Here...

That well-known raconteur of the early nineteenth century, Rev. Sidney Smith, was once heard to remark rather unkindly that 'I have, alas, only one illusion left, and that is the Archbishop of Canterbury.' Meteorologists are similarly impoverished. They have succeeded, by and large, in finding an explanation for all atmospheric phenomena, save one. There is still no rhyme or reason to *ball lightning*; indeed most professionals are sceptical that it exists at all. It remains one of the great enigmas of the science.

There are many anecdotal accounts of the appearance of a moving fire-ball during a thunderstorm. Although it is a very rare occurrence, it has been seen sufficiently often by reliable witnesses not to be dismissed entirely as hoax or mere imagination. It is usually described as a red or orange 'ball of fire', six to twelve inches in diameter, drifting along two or three feet above the ground. It lasts only a few seconds, and then disappears as quickly as it came, often after making contact with a metal object. This strange fire-ball has been known to occur both indoors and in the open air. It seems to be relatively harmless; 'lambent but innocuous', as another cleric once described the scintillations of a colleague's wit. People who have come into contact with it a have suffered no ill effects other than slight shock. The vast majority of occurrences take place in thundery conditions, when ordinary lightning has been observed as well.

But despite the hundreds of reports, scientists are still unsure if ball lightning really exists. Some attribute the phenomenon to 'brush discharges' of static electricity in the atmosphere; others assert that it is merely an optical illusion, caused by an after-image on the retina of the eye immediately following a lightning flash seen by the observer.

Two other factors increase their doubts. It appears that no professional observer of the weather has ever seen a fire-ball, despite their having collectively watched thousands of lightning flashes. And any statistical studies which have been carried out on reported sightings indicate that up to 80 per cent of the alleged occurrences may be readily explained by other means; of the remainder a high proportion can be called in question because of the time which has elapsed between event and report. Ball lightning remains the meteorological equivalent of Churchill's Russia: a riddle, wrapped in a mystery, inside an enigma.

A Foolish Flame

It was Ralph Waldo Emerson who remarked that 'a foolish consistency is the hobgoblin of little minds'. It is comforting advice for a meteorologist on those rare occasions when his forecasts are less than perfect, and he or she is obliged to start again from a fresh premise. But there are even closer atmospheric connections: in ancient folklore, a hobgoblin was a mischievous or evil spirit, and the name was commonly applied to one of nature's most intriguing mysteries, the will-o'-the-wisp.

Insofar as will-o'-the-wisp can be regarded as a scientific phenomenon, its proper name is ignis fatuus (the foolish fire). It is most commonly seen in bogs and graveyards, and a mild, wet, autumn night provides its favourite ambience. It is a tiny brightly coloured flame a few inches high, and is often described as resembling a lantern being carried by a person moving in a zigzag line, which explains its other name, the Jack o'Lantern. In Ireland our forebears called it *Liam an tSoip.*

The will-o'-the-wisp in folklore has a macabre reputation. In some cultures the dancing flames gliding in the silence of the night between the gravestones of a lonely churchyard were thought of as the souls of stillborn children, flitting between the upper and the lower worlds. Solitary travellers were lured to death as they followed the moving light across a treacherous marsh, 'leading men up and downe in a circle of absurditie, so they never knowe where they be.'

But will-o'-the-wisp is as mystifying in a natural sense as it is awesome in a supernatural context. It is believed to be caused by a flammable gas emerging from the ground, a gas capable of spontaneous combustion when exposed to air, and which burns without smell or smoke. Gases with these characteristics are sometimes a by-product of animal or vegetable decomposition, and the flame they produce on ignition is a chemical luminescence peculiar for its relatively low temperature.

No one has yet succeeded in capturing this elusive substance which ignites with such apparent ease. And such a *coup de gaz*, to coin a phrase, is now unlikely, since nowadays a will-o'-the-wisp is rarely seen. Some say that this is because conditions for its appearance are less ideal; bogs are drained and cultivated, and the ubiquity of artificial light might make it hard to see. But others say it never really happened anyway; that it was indeed the ignis fatuus, the fire of fools.

Chasing Rainbows

The eponymous rainbow becomes a symbol of hope for Ursula Brangwen in D.H. Lawrence's novel. Looking out disconsolately over the bleak dank dirty collieries of Cossethay she sees, all of a sudden – 'in the blowing clouds a band of faint iridescence touching a portion of the hill.'

> Steadily the colour gathered mysteriously from nowhere, and the arc bended and strengthened itself till it arched indomitable, making a great architecture of light and colour in the space of heaven, its pedestals luminous in the corruption of new houses on the low hill. She saw in the rainbow the earth's new architecture – the old, brittle corruption of houses and factories swept away, and the world built up in a living fabric of Truth, fitting to the over-arching heaven.

To the ancient Greeks the rainbow was the many-coloured robe of Iris, the winged messenger of the Gods; in biblical times it was the Sign of the Covenant, a reminder of the promises of Jehovah to Noah after the Flood; and to the Old Norsemen it was 'the bridge of the gods'. But to meteorologists the rainbow is a simple optical phenomenon, caused by the reflection of sunlight in raindrops.

To form a rainbow, millions of little raindrops act like tiny mirrors. Rather than being reflected on the surface, the light passes into each drop and is reflected from the back. This has two important consequences; firstly the optical properties of a sphere of water are such that the angle of reflection is always 42 degrees; and secondly, as it passes through the drop, the light is broken up into its constituent colours, giving the rainbow its familiar pattern with violet on the inside and red on the outer edge.

The 42-degree angle of reflection limits the occurrence of rainbows. None will appear unless the raindrops are located in such a way that the eye catches light reflected by them at precisely this angle. As a result, the bow is part of a circle which has an angular width of 42 degrees, and the sun is always *behind* the observer. The centre of this circle is at what is called the *anti-solar point*, a point on a continuation of an imaginary line joining the sun and the observer's eye.

The appearance of a rainbow depends very much on the time of the day. Since it is always on the opposite side of the sky from the sun, it follows that morning rainbows are in the west, evening rainbows in the east. Moreover when the sun is low in sky, the anti-solar point is nearly above the opposite horizon, so the rainbow stands out clearly as almost a full semicircle; conversely, when the sun is high, only a tiny part of the arc may be seen very low down in the sky. Indeed if the sun is higher than 42 degrees, a rainbow is impossible; and another impossibility in these parts is a rainbow to the south, for the simple reason that the sun never appears to the north.

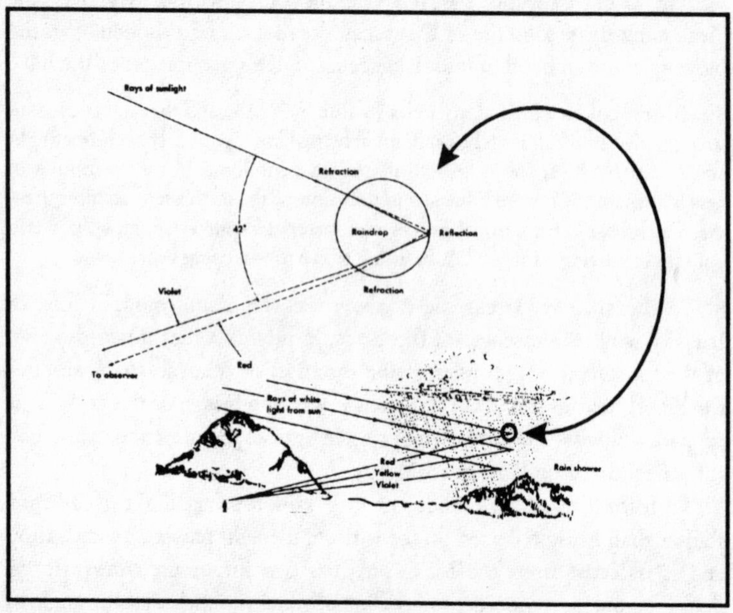

The optics of the rainbow. (Anne McWilliams, adapted from *Fundamentals of Meteorology* by Louis J.Battan)

Somewhere, Over the Rainbow ...

Rainbows often come in pairs. A secondary rainbow, concentric with, but larger than the main, is commonly regarded as exceptional; in fact, if you look closely, it is nearly always there. Somewhere, over the rainbow lies another rainbow!

The reason for the secondary bow is that when a ray of sunshine is reflected from the far side of a raindrop, not all the light then succeeds in making its escape. Some of it is reflected a second time from the inner surface of the sphere of water, and only emerges from the raindrop at its next attempt. The geometry of a transparent sphere is such that this twice-reflected ray of light finally emerges from the drop at an angle of 51 degrees to the original ray of sunlight. These rays which have experienced dual reflection give a second rainbow, well outside the first, with an angular size of 51 degrees.

Reflection – mature or otherwise – reverses everything. The colours of the rainbow are arranged in a certain way when the engendering light has suffered only one reflection. In the case of a second reflection, the positions of these colours are reversed. Thus a primary rainbow has violet on the inside and red on the outside; the secondary bow, a larger, fainter, optical echo of the first, has its colours arranged the other way around.

The secondary bow is fainter than the primary because most of the available light from the originating sunbeam has gone into the formation of the main rainbow; only a small residue is reflected a second time to give a secondary. In theory there are third- and fourth-order rainbows, resulting from multiple reflections of the ever-diminishing residue of light inside the drop. In practice not enough light remains to make them visible.

A Holy Sight

In religious artistic circles – or is it artistic religious – the terms 'nimbus', 'halo' and 'glory' are synonymous. Indeed I use the word 'circles' advisedly; all three are different names for the bright circular glow which in religious art traditionally surrounds the heads of the just and righteous. But confuse the three in the company of a weatherperson, and you are likely to be viewed disdainfully down the entire length of an elevated meteorological nose.

Nimbus is merely the Latin word for a cloud. It is appended, fore or aft, to many of the official names for clouds – like *cumulonimbus* or *nimbostratus.* A halo, too, is a relatively frequent phenomenon, being the large white luminous ring which sometimes surrounds the sun or moon when partially obscured by very thin high cirrus.

The glory is less common. For most of us the only opportunity we have of seeing it is when we travel by aeroplane, but in those circumstances it is rare not to see it. It is in evidence when the aircraft flies in sunlight above a layer of relatively low cloud consisting of water droplets, as distinct from the much colder and higher clouds composed of ice particles.

The glory is a series of concentric coloured rings arranged target-fashion around the shadow of the aircraft projected onto the cloud below. It is not in any sense caused by the plane's shadow; the glory and the shadow are two distinct phenomena, but it so happens that both occur in the same place: at the anti-solar point. They are centred at the spot on the cloud which lies on an extension of the line joining the sun and the aeroplane.

The glory is caused by a process called *diffraction.* The water droplets of the cloud interfere with the direct progress of the tiny waves of sunlight, and split the light into its constituent colours, the familiar colours of the spectrum. The light which is diffracted in this way to give the glory is sunlight which has already been re-directed back towards the aeroplane by internal reflection inside the droplets of the cloud.

Glories vary considerably in size, depending on the radius of the cloud droplets. Indeed as the aircraft proceeds along its track, the glory may expand and contract, as it re-acts to the changing composition of the cloud-layer below. The smaller the water droplets in the cloud, the larger the diameter of the glory.

An Eerie Eyrie

The Harz Mountains in northern Germany straddle the former Kingdom of Hanover and the equally former Grand Duchy of Brunswig, and lie close to the obsolete border between East and West. Their highest peak, slightly taller than our own Carrauntuohil in Co. Kerry, is called the *Brocken*; it is twice famous, the circumstances of its notoriety in each instance being somewhat spooky.

The older association is with Walpurgisnacht, the ancient pagan festival on the eve of May Day. St Walpurga herself seems to have been a harmless soul, who left her native England in the eighth century to try to convert the Germans to Christianity. Her only connection with Walpurgisnacht was that her feast day happened to fall on the eve of May 1st. Sinister events, however, took place on the Brocken on Walpurgisnacht; according to German legend, a witches' Sabbath took place there every year, and strange sights indeed awaited those who made the climb.

Strange sights are still to be seen upon the Brocken. 'The spectre of Brocken' is a simple optical phenomenon, but its appearance on the eponymous mountain was so common that it came to be associated with it. It can be seen here in Ireland, however, if the weather conditions are right. It consists of a giant shadowy figure looming from the early morning or late evening mist.

Despite its weird appearance, the Brocken spectre is easily explained: it is simply the observer's own shadow, cast by the sun when it is low in the sky onto a bank of cloud or mist. Its mystique lies in its unexpectedness, and in its unfamiliar shape.

The apparition has a strange triangular appearance. This is because, unlike most of the shadows we come across, it is not localized on a particular plane surface; it is projected *through* the mist, and extends over a depth of perhaps twenty or thirty yards. Rays of light just grazing the body of the observer, and forming the edge of the shadow, are subject to the same perspective effect as railway lines which appear to converge as they disappear into the distance.

The spectre of the Brocken is often seen on mountain tops, because low clouds often provide the 'screen' onto which the shadow is projected. Now and then it can be seen at ground level, if one happens to be in the right spot relative to the early morning sun and a nearby bank of fog.

Sunshades

We tend to think of the sky as blue, but in fact it can be any colour you care to mention; it can be, to quote Shelley, 'yellow and black and pale and hectic red', depending on the circumstances and the time of day. Most of the more dramatic variations happen when the sun is low in the sky, around sunrise or sunset.

The colours arise because of the effect of the atmosphere on the white light of the sun. This 'white' light is in fact a mixture of a whole range of colours, ranging from red, through yellow, green and blue, to violet. But if one or more of the constituent colours is filtered out for some reason, the remaining colours usually add up to something quite different from white. And the atmosphere, depending on its composition at a particular time, has precisely this effect.

The removal takes place by the process of *scattering*. Waves of any kind are often diverted from their original path by obstacles, just as a boulder interferes with water waves, sending wavelets off in many new directions. Tiny particles in the atmosphere interfere with the free passage of the rays of light through the air, and 'scatter' them in a different direction from that in which they were originally headed. The short wavelengths of blue light are relatively easily scattered; the longer wavelengths at the 'red end' of the spectrum are more resilient, and require larger particles to scatter them effectively.

When the sun is high in the sky it appears white, because little scattering has taken place, and its 'mixture' of colours reaches us almost intact. As the sun sinks near the horizon, the sunlight reaches us at a very oblique angle, and has to travel a much longer path through the atmosphere. This long path allows time for more scattering to take place, so that by the time the sunlight reaches us, its 'bluer' constituents – the violet, blue and green light – have been extracted, leaving only the more resilient red wavelengths to reach our eyes.

This, of course, brings us to the old saying: 'Red sky at night is the sailor's delight; red in the morning is the shepherd's warning.' Is there any truth in it? It depends on the shade of red. A moisture-laden atmosphere is very efficient at the process of scattering. Evening or morning sunlight which is a very deep shade of fiery-red has been greatly affected in this way, and suggests high moisture levels aloft, which augurs ill. But if the evening sky is a pale and delicate shade of yellowish-pink, it indicates a dry atmosphere, which of course is a good sign in the short term.

Hill Scene Blues

You haven't lived until you've read *A Shropshire Lad.* In those poems of A.E. Housman, the rural sentimentality of 'The Deserted Village' or Gray's 'Elegy' is laced with the sharpness of bitter-sweet experience, and seasoned with those hidden feelings for which only the Germans have found the right words, like *weltschmerz* and *schadenfreude.* 'What are those blue-remembered hills', he asks in one of them, 'what spires, what farms are those?' But let us return to meteorology, and first answer a quite different question: why were the hills *blue* in the first place?

On hazy summer days the distant hills are not the green or brown you might expect, but a definite shade of blue, and indeed the further away you happen to be, the more pronounced this bluish tinge. As in the case of the blue sky, the explanation lies in the scattering of light waves by tiny particles suspended in the air.

Tiny dust particles in the air, and indeed the very molecules of the air itself, are efficient scatterers of the shortest light waves: those at the *blue* end of the spectrum. As a consequence, the blue light in a ray of sunshine is diverted in many different directions as it passes through the atmosphere. Now as you look at a mountain in the distance you see, naturally enough, the mountain itself. But superimposed on this image is another source of light; some of the blue from rays of sunlight passing through the air is diverted in your direction by the process of scattering. It is this scattered blue light which gives the scene its characteristic colouring.

A number of other curiosities can be explained once you know what is going on. The farther away the hills, for example, the more air there is between you and them to scatter light in your direction, and the bluer they will appear. Moreover, in dry settled summer conditions, when the barometer is high, the number of impurities in the air is higher than usual, allowing a greater degree of scattering and enhancing the bluish tinge. After a few heavy showers, when the rain has washed away the dust, the air becomes clear and transparent and the natural colours of the distant hills are unshrouded by any veil of blue.

A Terrible Beauty

These are a few lines written by Ettie French, daughter of the famous Percy:

> My father was thirty when he was first attracted to landscape painting. He was enjoying a life full of congenial activities and deeply in love with the girl he was going to marry, when a series of wonderful sunsets over Lough Sheelin completely bowled him over. He went out every evening and tried to capture in paint the colours, which were due to volcanic dust.

Percy French was thirty in the year 1884. The volcano in question was on Krakatoa, a small island in the Sunda Straits, between Java and Sumatra in what is now called Indonesia, and in the same part of the world as the recently erupted Pinatubo. Krakatoa erupted on 28 August 1883 and the massive explosion was heard over three thousand miles away; an estimated four cubic miles of debris was hurled into the atmosphere in a few frenzied hours.

The immediate impact was to reduce the surrounding area to almost total darkness for some days. But there were more lethal consequences. Giant tidal waves reached heights of 120 ft and traversed the entire Indian Ocean devastating everything in their path; the official death toll came to 36,417. When the eruption had ended, only one third of the island of Krakatoa remained above water, and scores of new islands of steaming pumice and ash lay where previously the sea had been over one hundred feet deep.

For those not in the immediate vicinity there were positive aspects to this spectacular event. From a meteorological viewpoint the eruption of Krakatoa was significant, since close monitoring by scientists of the volcanic dust in its aftermath contributed greatly to contemporary understanding of circulation patterns in the high atmosphere.

And there was also a visual impact. Tiny particles of ash penetrated the stratosphere to heights of thirty miles or more around the entire globe, and caused the sun to appear blue and green in many parts of the tropics. The volcanic dust also gave spectacular sunsets right around the world during the succeeding two years, and started Percy French on his subsidiary career as a talented water-colour artist.

Songwriter Percy French (1854–1920) took up landscape painting because of the spectacular sunsets following the eruption of Krakatoa in 1883.
(From *A Picture of Percy French* by Alan Tongue)

A Monster Inspiration

Mary Wollstonecraft Shelley was born on 30 August 1797. At the tender age of seventeen she eloped with future husband Percy, and spent the next eight years doing the grand tour in the company of the poet and his disreputable friends. In the summer of 1816, she found herself in Switzerland, on the shores of Lake Geneva, with Shelley, Byron, and the writer Polidori. Now the summer of 1816 was a dreary one throughout the whole of Europe. Mary herself described what happened:

> At first we spent our pleasant hours upon the lake, or wandering on its shores. But it proved a wet and uncongenial summer, and the incessant rain often confined us to the house for days on end. 'We will each write a ghost story', said Lord Byron.

And so they did! The efforts of Shelley and Byron were quickly abandoned, but Polidori produced a story called *The Vampire.* And Mary Shelley's tale became a classic; it was published in 1818 under the now-famous title *Frankenstein,* a bizarre Gothic novel about a man-made monster which turns on its creator when he gives to it the spark of life.

But it is when we analyse the reasons for the appalling weather which provoked such creativity that we return again to a volcano. Just over a year before, on 7 April 1815, Mount Tambora in the East Indies had exploded. Thirty-six cubic miles of rock and dust were hurled into the upper atmosphere, and twelve months later the dust in the stratosphere had spread into a world-wide veil. It reduced the penetration of the sun's rays, caused a drop in average global temperatures, and distorted wind patterns throughout the world.

Prevailing winds to a large extent dictate the weather, and the change brought strange anomalies. Regions usually wet became almost arid for a time, and dry areas experienced more than their fair share of rain. Most of the Continent that year was cold and wet; indeed throughout Europe and North America conditions were so unusual that the year 1816 is remembered as 'the year without a summer'.

Bright Rays

The average meteorologist has a great weakness for multi-syllabic epithets. Rather than follow the advice of Ben Johnson and 'be not retrograde, but boldly nominate a spade a spade', he prefers to use a long word if he can find one. That which you might call a sunbeam, he will tell you is a *crepuscular ray*. A crepuscular ray occurs when a beam of sunlight passes through a gap in the clouds, and the contrast between brightness and shadow provides a well-defined shaft of light. The same phenomenon can sometimes be observed in a darkened room, when sunshine streams through a single window, or through a gap between the curtains, its path being clearly illuminated by the dust hanging in the air. Out of doors, the path for the isolated beams is provided by chance openings in a layer of cloud.

When the sun is high in the sky, particularly in showery weather, crepuscular rays can sometimes be seen in the distance piercing a layer of relatively low cloud. They are directed downwards at a slight angle to the vertical, and stand out clearly because the light is scattered by mist and dust suspended in the atmosphere. It used to be said in these circumstances that 'The sun is drawing water'; it was popularly, but of course incorrectly, supposed that the rays were concentrations of water-vapour, allowing water to return to the sky so that it could rain again.

But it is when the sun is low in the sky, or even below the horizon in the late evening or early morning, that crepuscular rays can be seen at their most spectacular. The red or rose-coloured sunbeams diverge upwards from behind a distant cumulus or cumulonimbus, spreading their crimson stripes of light and shadow in a huge fan, and providing the vista so admired by generations of seascape painters. Indeed it is from its striking appearance in this particular guise that the phenomenon gets its name. *Crepusculum* is the Latin for 'twilight'; they are literally 'twilight rays'.

The fan-like shape, of course, is simply an optical illusion. The light comes from the sun, which is so far away from us that its rays are as nearly parallel to each other as makes no difference. They appear to converge near the horizon for the same reason that straight railway lines, or the two sides of a long wide street on which you might happen to be standing, appear to meet in the very far distance.

Normal Visions

'A damsel with a dulcimer, in a vision once I saw', recorded the hung-over Samuel Taylor Coleridge. Few of us can claim to have had so exotic an experience, but in very hot weather, visions of a more mundane variety are commonplace. All of us, from time to time, have seen a *mirage*.

A mirage is an optical illusion, a distorted view of the surrounding world brought about by the eccentricities of a non-uniform atmosphere. The best-known examples occur when there are very large changes in air temperature over a short vertical distance; rays of light passing from one layer of air to another at a different temperature are bent, or *refracted*.

Any object illuminated by the sun – or by anything else for that matter – sends out rays of light in all directions. These rays, as a general rule, travel in straight lines. The eye of an observer looking at this object picks up the rays heading towards it, and the brain, well-attuned to rectilinear propagation, estimates the position of the object from experience. But when refraction takes place, the eye is deceived.

Refraction often occurs to a marked extent in hot weather over a concrete or tarmacadam surface. A very thin layer of air in immediate contact with the ground becomes very warm indeed, and there may be a drop in temperature of 15 degrees or more in the first half inch above the surface. Viewing, for example, a distant tree in these circumstances, we can see it first of all in the normal way; but in addition, some of the rays of light proceeding groundwards from the tree are bent upwards again by refraction, and reach our eyes from a quite different angle.

The eye becomes bamboozled; it perceives two quite different versions of the same tree: the normal one, and a mirror image of the tree inverted underneath. In the same way we can sometimes observe inverted images of the sky or of a distant landscape. The optical effect is such that a hot road appears to be covered with pools of water, which grow larger and clearer if one stoops to look at them, and which appear to reflect bright and coloured objects in the distance. But what we take for water is nothing more than hot air, and what we see is an apparent reflection of the clear blue sky above. The conventional idea of a mirage – that which has traditionally tantalized the thirsty desert traveller – is a rather more complex phenomenon, but the basic optical principle involved is just the same.

VII *Forecasters and Forecasting*

Pepys at the Weather

To the man in the street of seventeenth-century London, weather was a daily concern. Rain, snow and frost often made the ways 'very foul', while in the summer it was 'so dusty that one durst not breathe'. Even on the river Thames, which was still London's main highway, there was no escape; storm, tide, mist and darkness hindered passage by boat and the watermen would sometimes become completely lost.

A keen observer of these conditions was Samuel Pepys, who faithfully recorded a daily weather bulletin in his famous *Diary*. On 21 January 1661, for example, he was wondering if something funny was happening to London's climate: 'It is strange what weather we have had all this winter; no cold at all, but the ways are dusty, and the flyes fly up and down, and the rose bushes are full of leaves, such a time of the year as was never known in this world before here.'

Pepys was the son of a London tailor, but he had family associations with the influential Earl of Sandwich, who helped Pepys throughout his career. With the aid of his patron, combined with his own natural ability, Pepys in due course became Master of Trinity House, was twice a Member of Parliament, and was for two years President of the Royal Society. Over the years he became very prosperous, and died a wealthy man.

Pepys began his *Diary* on 1 January 1660, when he was twenty-seven years old. He kept it faithfully for almost a decade, writing in shorthand to ensure privacy, with the result that its contents remained a mystery until 1825. It documents not only his own escapades and the life and manners of the times, but also the daily sequence of the weather in London during the 1660s. Although no meteorologist, Pepys had a keen interest in the subject, his visual observations being aided by 'a very pretty weather-glasse for heat and cold' which he had received as a present in 1663.

Pepys continued writing until 31 May 1669, when fears for his failing eyesight caused him to bring the *Diary* to an end. His closing words were appropriately sombre: 'And so I betake myself to that course, which is almost as much as to see myself to go to my grave – for which, and all the discomforts that will accompany my being blind, the good God prepare me!' In fact, Pepys lived for another thirty-four years, and died at Clapham on 26 May 1703.

An Historical Black-Spot

There is an old Irish saying to the effect that only two things in this world are too serious to joke about: potatoes and matrimony. It reflects the importance of both, not necessarily in this order, to indigenous Irish life.

Nobody knows quite when matrimony hit our shores, but the arrival of the humble spud is generally put at around 1585. By the mid-nineteenth century, the potato had become the staple diet for the vast majority of Ireland's population. If anything were to happen to cause a failure of the crop, the result would be disaster. And that, as we know, is precisely what happened in 1845, with the arrival of the Great Irish Famine.

The weather was largely responsible for this great tragedy. Although a great many social and political factors contributed to the strange and terrible events of 1845 and thereafter, the cause was the prevalence of a specific combination of meteorological parameters. The weather during those years was ideal for the spread of potato blight, a fungal disease which thrives only in very specific atmospheric conditions.

Potato blight spreads only when high temperature and very high relative humidity coincide. For the spores to infect the plants three conditions must coincide over an extended period: the leaves must be wet, the relative humidity must exceed 90 per cent, and the temperature must be greater than 10°C.

Such weather was a frequent visitor to northern Europe in the early summer of 1845. The blight first appeared in Belgium in late June, and by mid-August it had spread to northern France and the south of England. By September 1845 it had reached Ireland. On continental Europe and in Britain the following summers were dry and hot, so the blight died out but in Ireland the weather was abnormally wet and warm. The blight thrived, and the rest, as they say, is history.

Weather Worms

One, can, of course, just listen to the weather forecast, but, as Charles Kingsley remarked in *Westward Ho!*, 'There are more ways of killing a cat than choking her with cream.' And there are more exciting ways of weather-watching. Some deduce the weather trends by watching seaweed, or by careful note of rheumatic joints or corns. The most bizarre predictor, however, must surely be the medicinal leech, whose prowess in this regard was well known in the last century.

The medicinal leech is an aquatic blood-sucking worm, which by some evolutionary legerdemain has succeeded in acquiring three sets of jaws. It was widely used in olden days to relieve the sick of any blood which the medical profession at the time believed to be superfluous. But it is not its usefulness as a medicinal aide that concerns us here; it is the reputation of the leech as a weather forecaster *par excellence.*

One puts one's weather leech in a bottle of water. Leeches, apparently, relax at the bottom of their bottle during fine calm weather, but half a day or so before a change, the worms begin to turn; they move steadily upwards towards the surface. If rain is at hand they move out of the water altogether, sticking to the glass at the side of the neck. And if he expects a storm, the leech will cleverly curl himself into a ball, and wait like that until the storm has gone. Once the weather settles down, the leech drifts slowly down again to the bottom of his bottle, ready once more to leap into action at the slightest hint of change.

The appropriately named Dr Merriweather, a medical practitioner in the Yorkshire town of Whitby in the middle of the last century, became a recognized expert in this somewhat bizarre method of weather prediction. He succeeded in building an apparatus by means of which a leech confined in a bottle of water rang a little bell when a storm was expected. He displayed this unique invention at the Great Exhibition in the Crystal Palace in London in 1851, and advised the government to establish leech warning stations around the English coast.

It must be recorded, alas, that the government of the day did not see fit to act upon the advice of the good Doctor. But his expertise was not entirely lost to posterity. Merriweather published a pamphlet on the subject which was published in the same year, entitled 'An Essay Explanatory of the Tempest Prognosticator'.

An Act of Hatfield

Forecasting the weather is all very well, but the real trick is to *control* it, to turn the rain on when you want it, and even more importantly, to turn it off when you don't! We know, of course, that the amounts of energy involved in the atmospheric processes are so enormous that it would be impossible to provide in any controlled way sufficient energy to effect *direct* change. All that one might hope to do is to give nature a slight push; to add that essential extra ingredient to initiate a process which was just on the point of happening anyway, but which might not have! That is precisely what the most famous rainmaker in history, Charles Hatfield, claimed to do.

Hatfield was born at Fort Scott, Kansas, in 1875. He began his rainmaking career in 1905, having, he claimed, been a student of meteorology for seven years. His method was both impressive and enigmatic: he would set up a number of large towers in the vicinity where rain was required, each one surmounted by a vast vat of boiling liquid to which he added secret chemicals at prescribed intervals. He did not claim that this concoction brought the rain; he merely attracted the clouds, he said, and the rain followed!

Hatfield had a number of presumably coincidental but widely publicized successes, and as result his expertise became widely known. The US Weather Bureau issued press releases to the effect that he was a fraud; this did not diminish his popularity, and he made a great deal of money on lecture circuits. But his *cause célèbre* was San Diego, California.

After a number of rainless months in 1915 the city fathers of San Diego became painfully aware that their reservoirs were empty. They sent for Hatfield, who agreed to fill their reservoirs to overflowing for a modest $10,000. He set up his towers, and began rainmaking on 1 January 1916 – and the rain duly came. But it came, and came, and came again! There was widespread flooding in the ensuing weeks, houses were washed away, and on 27 January the Otay Dam collapsed with the loss of many lives.

Understandably, the city council blamed Hatfield. They sued him, and the case went all the way to the Californian Supreme Court, which ruled that the rain was an 'act of God', and that Hatfield had nothing to do with the disasters. There was only one consolation for the San Diego city councillors; since the rain was declared by law to be an act of God, it was obviously not an act of Hatfield and they withheld his $10,000 fee.

(From *Scientific American*, 27 November 1880)

Murphy's Winter

'One crowded hour of glorious life is worth an age without a name', according to the poet Thomas Mordaunt. Patrick Murphy's crowded hour arrived on 20 January 1838. It made him, for a little time at any rate, a very wealthy man.

Murphy was an English dilettante of the sciences. He was prolific, and his published works included books and papers on the force of gravity, on electricity, and on the solar system, and two pedestrian works on meteorology. All except one, however, are forgotten. Like many eighteenth- and early nineteenth-century meteorologists, Murphy was convinced that weather changes were related to the moon, to other planets and even animal behaviour, and in 1837 he produced his *magnum opus.* In that year P. Murphy Esq., MNS (which proudly stood for 'Member of No Society') published *The Weather Almanac (on Scientific Principles, showing the State of the Weather for Every Day of the Year of 1838).*

This somewhat presumptuously titled volume quickly became a runaway best-seller. Its success was due entirely to the entry for 20 January 1838, under which date he simply wrote: 'Fair, and probably the lowest degree of winter temperature.' But by happy chance it was indeed the coldest day in living memory; the thermometer in London dropped to –20°C, the Thames was frozen over, and a sheep was roasted whole upon the ice at Hammersmith.

Murphy became famous overnight. His *Almanac* ran to forty-five editions, and customers mobbed his publishers to obtain their copies of the book. Its author made a profit of £3000 – a considerable sum in those days – and the season ever afterwards was known as 'Murphy's Winter'.

But Murphy's hour was brief. Almost immediately, it seems, he lost his windfall on an unsuccessful speculation on the corn market. After the initial riot of success, the sales of the *Almanac* languished as it became evident that no faith could be placed in its prognostications. He tried again in 1839 and in various other years until his death in 1847, but the sales were very limited indeed. In due course, what could have been his epitaph was penned by a detractor to *The Times,*

> When Murphy says 'frost', then it will snow,
> The wind's fast asleep when he tells us 'twill blow,
> For his rain we get sunshine, for high we have low.
> Yet he swears he's infallible – weather or no!

Taking Umbrage

An unusual feature of Daniel Defoe's *Robinson Crusoe* was the fact our hero equipped himself with an umbrella for protection from the sun. The story was written in 1719, at a time when no self-respecting male would have dared to carry such an implement. The first man reputed to have done so was Jonas Hanway, a Londoner who started the fashion around 1750 and who received a hard time from the populace of his native city before his very practical idea gained limited acceptability more than thirty years later.

There was, however, nothing new about umbrellas themselves. In the guise of the parasol, intended to protect its owner from the sun, the idea is a very ancient one. These portable sunshades are believed to have first appeared in China many centuries BC, at which time they had great ceremonial significance as an emblem of exalted rank. Often of great diameter, with many tiers adorned with elaborate fringes and tassels, they were carried high above the heads of civil and religious dignitaries.

In later times parasols were used by the women of Greece and Rome to protect them from the unflattering effects of the midday sun. Their use died out during the less genteel era of the Dark Ages, but they were revived by women of fashion in southern Europe during the Renaissance and are believed to have appeared in England just before 1600; during the seventeenth and eighteenth centuries, the umbrella or parasol again became an essential accessory for any well-dressed lady.

It was at this stage that Jonas Hanway had his bright idea, and his idiosyncratic use of the umbrella to keep off the rain caused great hilarity. But his behaviour provoked an angry reaction in some quarters. Coach owners claimed it would ruin their trade. Others thought the umbrella was an insult to God; unless the rain were intended to wet people, they argued, it would never have been sent in the first place; certainly no one had the right to keep off the rain with an umbrella!

It took many years for this strange accoutrement to evolve into common use. Many years after Hanway's death there was still only one umbrella in the whole city of Cambridge, kept at a shop and hired out like a cab by the hour. In due course owners of inns and coffee houses began to keep one which could be borrowed by customers going to and from their carriages, and by the end of the nineteenth century the umbrella had become an indispensable personal protector against the whims and sorrows of our outrageous climate.

The Parson's Tale

Once upon a time a lecturer in mathematics from Oxford University took his three nieces for a boat trip on the river Isis. Rev. Charles Dodgson, better known to succeeding generations as Lewis Carroll, was thirty-two at the time; his eldest charge, Alice, was ten. The quartet rowed from Oxford to Godstow and back, and in the course of the journey *Alice's Adventures in Wonderland* were related for the first time.

Alice Liddell, for that was the young lady's name, remembered the day well in later life:

> I believe the beginning of Alice was told one summer afternoon when the sun was so burning that we had landed in the meadows down the river, deserting the boat to take refuge in the only bit of shade to be found, which was under a new-made hayrick. Here from all three came the old petition of 'Tell us a story', and so began the ever delightful tale.

Dodgson, too, recorded the events of the day in his diary. Thus they started one of the great controversies of modern literature.

The day in question was 4 July 1862. The controversy concerned the weather on that day. Alice asserted that the day was warm and sunny but the critics had their doubts. Official weather records appeared to show that a good deal of rain fell in the twelve hours prior to 2 a.m. on 5 July 1863. They concluded that the afternoon of the 4th must have been cool and rather wet, and that the memories of those present on that famous occasion must have been playing them false. Curiouser and curiouser!

This controversy was settled some years ago by an Irish meteorologist, the late H.B. Doherty. He discovered that archived issues of *The Times* contained detailed weather reports from many of the sea-ports around Britain and Ireland, and, using these as one might use modern weather observations, he constructed weather-maps for the period in question.

These showed that an active front had passed the Oxford area in the early morning of 4 July, and that another moved in from the west late that evening. In between, however, at the time of the boat trip, a transient ridge of high pressure gave the idyllic conditions so fondly recollected by the main participants.

> Thus grew the tale of Wonderland,
> Thus slowly, one by one,
> Its quaint events were hammered out –
> And now the tale is done!

The Bane of Spain

Towards the end of May 1588 there was great excitement in the port of Lisbon. The Spanish Armada was about to sail! The vast fleet of 130 ships assembled by King Philip II to challenge the might of Elizabeth's England had been ready for some weeks, but unfavourable winds had delayed its departure. Its problems with the weather were only beginning.

The plan was simple. Having negotiated the English Channel, it was proposed to take on board an army of 30,000 men assembled in the Spanish Netherlands under the command of the Duke of Parma. From there the invasion of England would be launched. Everyone – even the English – was sure it would succeed. As it happened, the non-event is a prime example of a turning point in history in which the weather played a vital role. The weather was in a funny mood around that time. The first half of the sixteenth century, by and large, had been a period of genial climatic conditions over much of Europe, and average temperatures were higher then they had been for some time. But by the 1580s, the picture had changed dramatically. Temperatures were significantly lower, and the summers were colder, wetter and much windier than they had been in the early years of the century. Even William Shakespeare noticed it, he wrote in *A Midsummer Night's Dream* in 1594:

> The seasons alter: the spring the summer,
> The chiding Autumn, angry winter, change
> Their wonted liveries, and the mazed world,
> By their increase, now knows not which is which.

In this context Philip launched his ill-fated attack on England. Once out in the Atlantic, the fleet encountered further hindrance from the unfriendly elements, so it was early August before the Armada finally reached English waters. The military set backs it encountered there were relatively minor, but nonetheless sufficient to persuade the Spaniards to sail for home, heading around the north of Scotland and down the west coast of Ireland. But the weather had more unpleasant surprises in store.

A series of violent storms encountered off the Irish coast in mid-August proved to be their worst adversary yet. Twenty-five ships were wrecked between the Giant's Causeway and the Blasket Sound, and with them were lost many of the highest-ranking noblemen of Spain. Of the 130 ships that sailed on 30 May 1588, only slightly over half returned home that autumn.

Faithful Forecasts

In ancient India a Brahmin was a member of the priestly class. His function was to study and to teach the Veda – the sacred literature of the Hindu – and to perform ritual sacrifices. He also had another sacred task: such was the importance to the area of the rains brought about by the annual summer monsoon, that the Brahmin was required to forecast their arrival. Error incurred severe penalty; a false prophet was obliged to observe total silence for the rest of his life.

Our Western tradition had less faith in ecclesiastical weather forecasters. The attitude of our forebears was summed up in the rhyme:

> Better it is to rise betime,
> And make the hay while the sun doth shine,
> Than to believe in tales and lies
> Which idle friars and monks devise.

But if they ignored the clergy, our ancestors placed great store on the day of the week as a weather indicator. Some, for example, believed that: 'On Thursday at three, look out and you'll see what Friday will be.' Others were of the view that: 'If sunset on Sunday is cloudy, it will rain before Wednesday.'

In more modern times meteorologists have tried in a scientific way to see if there is, in fact, any detectable difference between the days of the week as regards weather. At first sight, there should be none, because the seven-day week is an entirely arbitrary division of time, unknown to nature and introduced by man purely for his own convenience.

An Englishman called Ashworth, however, spent many happy years in the 1920s analysing detailed rainfall statistics for Rochdale in Lancashire. He found that Sunday was the day of the week with the least rain, and also discovered that on weekdays there was more rain during working hours than there was in the evenings or at night but on Sundays the reverse was the case.

It was concluded at the time, that the pattern was a consequence of industrialization. It reflected the abundance during factory working hours of the tiny particles which facilitate the condensation of water vapour in the atmosphere. This, it was explained, might lead to more clouds, and hence to a greater frequency of rain. Perhaps some days are worse than others after all!

The Mayfly Mystery

Angling, according to that great seventeenth-century champion of the art, Izaak Walton, is 'a rest to the mind, a cheerer of spirits, a diverter of sadness, a calmer of unquiet thoughts, a moderator of passions, and a procurer of contentedness which begets habits of peace and patience in those that profess and practice it'.

The mayfly is a large insect with a 2-inch wing-span, and yellow diaphanous wings. Trout, it seems, find the mayfly quite irresistible, and jump excitedly from the rivers and lakes in their eagerness to catch it. Their frequently undiscerning enthusiasm is a golden opportunity for the angler, and makes the period when the mayfly is active a most fruitful time to go fishing.

These little insects spend all their larval lives underwater, and emerge into the air only for one short but final fling, a dance of death to provide for the continuation of their species. Each male disports himself with excited rapture among a thousand rivals for the favours of a single female, who is a rare prize among so many courtiers. Pond and stream and sluggish river come to life with this short-lived *danse macabre* – rarely does an individual mayfly survive more than a day or two.

The brief dramatic sojourn above water of these insects takes place in a short two- or three-week period in the spring. The date of the 'rise' is known to be related in a general way to the weather, but meteorologists have had little success in trying to devise a scheme to predict this fishy 'happy hour'.

It is known that mild sunny weather in the spring means an early rise. Efforts to relate it to more specific parameters have not been successful. It might, for instance be related to the lake or river temperature; it could depend on the amount of sunshine, since the solar radiation might heat the silt of the shallow waters where the larvae develop or indeed the sunshine may help the growth of the algae, which are food for the larvae, and thus bring about an early hatching. Or it may be connected with rainfall, since low rainfall means lower water levels, and therefore water richer in nourishment. The solution may lie in all, or some, of these factors, but so far the right formula has eluded all who seek it.

Showers of Blood

Every so often there occurs in Ireland a strange weather phenomenon which never fails to attract widespread attention. The populace wakes up one morning to find cars and other belongings covered in a fine red dust; they wonder if it came from outer space, and whether or not it is radioactive. A major occurrence of this kind took place, for example, on 29 November 1979.

We are by no means the first race or generation to be mystified by this scarlet visitation. Homer, apparently, relates how a 'shower of blood' fell upon the heroes of ancient Greece as a harbinger of death; Plutrarch speaks of showers of blood after great battles in the Cimbric War; and in AD 1117 in Lombardy the showers of blood caused such consternation among the locals that a meeting, not of meteorologists, but of Bishops was held afterwards in Milan to consider their origin. In the period 1800 to 1873 there were over twenty reported instances of showers of blood in Europe.

The explanation is relatively simple. The fine dust comes from North Africa, from the red soil of the Sahara. But for it to be deposited directly on our little island, a series of meteorological coincidences is needed.

Sandstorms in the Sahara are the first requirement. The reddish-brown particles of dust raised by these storms are carried high into the atmosphere by the powerful convective currents common in those latitudes. Once aloft, the lighter grains remain suspended and are carried along at high levels typically two to four miles above the ground.

For Ireland to be affected, the most favourable weather pattern exists when an anticyclone – or area of high pressure – lies over Western Europe. This gives easterly winds over North Africa, bringing the air-borne dust first westwards out into the Atlantic, and then north-eastwards to reach Ireland three or four days later.

Finally comes the last link in the chain. The airborne particles are too light, in general, to fall to the ground of their own accord. But if rain originates in the atmosphere at a higher level than the dust, many of the suspended red particles are washed to earth, carried downwards with the raindrops. The result is what our ancestors called a 'shower of blood'.

Meddlesome Mendicants

Shortly after his unforgettable revelation that all of Gaul was divided into three parts, Julius Caesar in *De Bello Gallico* made another very astute observation, *Homines id quod volunt credunt* (Men readily believe that which they wish to believe). This, combined with a cynical shrug of the classical shoulder, sums up the attitude of meteorologists to the St Swithin legend and weather lore associated with many other saints.

Swithin's interfering habits are well known: rain on the feast of St Swithin condemns us to rain on each of the following forty days. The saint's stubborn streak which gave rise to this tradition is also well documented. The young Swithin was a devout but not unambitious monk of Wessex in the south of England. He gained the favour of Egbert, King of Wessex, who entrusted to him the education of his son Ethelwulf. And when Ethelwulf in due course became king himself, he did not forget his old mentor; in AD 836 he appointed Swithin to be Lord Bishop of Winchester, an influential and very rewarding position at that time.

Swithin, however, did not allow his good fortune to turn his head. In fact he was a good bishop renowned for his humility; when he gave a banquet he invited the poor and lowly to his table, but not the rich and noble. And on his death in 862 he had arranged for his burial outside his cathedral 'in a vile and unworthy place', where water from the eaves might fall upon his grave. There he lay for more than one hundred years.

Swithin's reputation grew with the passage of time, and miracles were said to be wrought at his humble resting place. With his waxing fame, and no doubt growing importance as a lucrative diocesan asset, the good monks came to think of it as scandalous that the remains of such a holy man should rest in such an unseemly spot. They prepared to move the saint with solemn ceremony inside the cathedral, to a gold shrine purpose-built, and richly ornamented with precious stones. The day appointed was 15 July 971.

According to legend a tempest raged on that fateful day and the removal was postponed. And the next day it rained again. For forty days and forty nights it rained heavily and without intermission, until finally the attempted move was abandoned. Realizing the folly of disobeying the instructions of the holy saint, the monks left him where he was, and erected a simple chapel over his grave instead.

But Swithin is not alone. The superstition has its counterpart in several other European countries, although the operative saint, the tradition or myth attaching to him or her, and the date of their supposed taking charge of the weather, differ in each case.

In Scotland, for example, it is St Martin Bullion who controls the weather for forty days from 4 July; in France, Benedict, Medard, Protase, and Anne are all credited with Swithinian tendencies; the Belgians' rainy saint is Godelieve, whose traditional date is 27 July, while Italy's Swithin is St Bartholomew; the German's, on the other hand, look to the Seven Sleepers of Ephesus.

The interesting thing about all these superstitions is that while the supposed dates of commencement may differ, the period of influence is consistently forty days, a suggestion that all these ancient fancies had a common origin in the world-wide tradition of Noah's flood.

Cat-Casting

Over the centuries there has been a persistent belief that some animals have an instinctive feeling for the weather, that they can sense a change long before it becomes apparent by other signs. The belief is not unreasonable. To many species in their natural state, the weather is of great importance; a sudden storm may at the very least deprive them of their next meal, and may in certain cases be life-threatening. Nature may well have endowed them with a special sensitivity to atmospheric changes. But to cats, by all accounts, Nature has been very kind indeed!

When your cat sneezes she is telling you that it is about to rain. It may also mean rain if she begins to wash her face with her paws; Erasmus Darwin noted the phenomenon thus, among various other signs of poor weather on the way: 'Puss on the hearth with velvet paws/Sits wiping o'er her whiskered jaws.'

But cat-casting is no pursuit for the careless dilettante; the signs must be watched very closely indeed, because if the cat puts her paws above her ears during this operation, it augurs something quite different. As another sage described it – one whose powers of observation, one hopes, were significantly better than his ability to turn out rhyming couplets – 'If the cat washes her face o'er the ear/'Tis a sign the weather will be fine and clear.'

Our feline friends, however, have other meteorological talents besides the ability to forecast rain. If, for example, your cat rubs herself against a wall, or scratches against a post, it is apparently her version of a severe gale warning. And she is also adept at temperature prediction: in very cold weather, we are told, cats wash their faces before a thaw, and sit with their backs to the fire before a bout of snow. But when the rain or snow is finally over, the cardinal point to which your cat turns to wash its face shows the quarter from which the wind will very shortly begin to blow.

The Bloomsday Book

The day began as we all know when 'Stately, plump Buck Mulligan came from the stairhead, bearing a bowl of lather on which a mirror and a razor lay crossed.' It was Bloomsday, 16 June 1904, probably the most famous day in literary history. But what is known about its meteorology?

There was, that morning, 'warm sunshine merrying over the sea'. It was not completely calm, because Buck Mulligan's 'yellow dressing gown, ungirdled, was sustained gently behind him by the mild morning air'. And neither was the sky completely blue: 'A cloud began to cover the sun slowly, shadowing the bay in deeper green.'

There had been little rain in recent weeks: 'After hard drought, please God, rained, a bargeman coming in by water a fifty mile or thereabout with turf saying the seed won't sprout, fields athirst, very sad coloured and stunk mightily.' As the day matured, the breeze became obtrusive: '...the wind had freshened, paler, firm and prudent ... the whitemaned seahorses, champing, brightwindbridled, steeds of Mananaan'. And through it all, the sun shone fitfully, and it was warm and dank and sultry, humid.

Later, as Bloom walked past the Provost's house at Trinity, 'the sun freed itself slowly and lit glints of light among the silverware in Walter Sexton's window opposite'. Even by nightfall the weather still held good: 'The summer evening had begun to fold the world in its mysterious embrace. Far away in the west the sun was setting and the last glow of the all too fleeting day lingered lovingly on sea and strand.'

Then came the thunderstorm: 'By and by, this evening after sundown, the wind sitting in the west, biggish swollen clouds to be seen as the night increased and the weatherwise poring up at them and some sheet lightnings, past ten of the clock, one great stroke with a long thunder.' Thus ended Bloomsday, with a 'black crack of noise in the street here, alack, bawled, back. Loud on left Thor thundered: in anger awful the hammerhurler'.

What was it like on Bloomsday? The weather-chart for 16 June 1904 shows a deep depression centred between Ireland and Iceland, with a brisk, mild, humid southwesterly airflow over Dublin. It would have been rather windy on that day, broken skies allowing the sun just now and then to 'fling spangles, dancing coins'. Rain was approaching Dublin from the west, and who is to say there was no thunderstorm?

Halcyon Days

'Expect Saint Martin's Summer, halcyon days', cried Joan of Arc, as she bravely led her army to raise the siege of Orléans. The Maid was anticipating, of course, a period of calm and tranquillity in an otherwise war-torn and turbulent era.

The term 'halcyon days' still has a somewhat similar meaning. We think of a time long ago when all was well, a time of youth, peace and prosperity, gone probably never to return. But the original *halkyon days* were much more specific. Just as we have our Buchan Spells, St Swithin's Day, and Scairbhin na gCuac, the ancient Greeks, too, had their own weather lore.

Winter in Greece may be a good deal less severe than here, but it can be unpleasant enough nonetheless. Those shrewd weather observers of ancient times, however, noticed that a spell of sunny, mild and nearly windless conditions frequently occurred in mid-winter. This unseasonal period turned up with remarkable regularity around the winter solstice, and could last anything from two days to a few weeks. But very often it lasted for about fourteen days.

Now, it so happens that there was a bird common in Greece at that time called the *halkyon*, a species of kingfisher. It was noted that the halkyon laid and hatched its eggs in the period around the winter solstice, usually, in fact, in the period from seven days before the solstice to seven days after it. And this brief domestic urge of the halkyon coincided with remarkable regularity with the fine spell of weather which relieved the not-so-harsh Greek winter.

What was more natural than that the ancient Greeks should connect the two? The spell of fine weather became known as the *halkyon days*. The legend was that the halkyon spent the first seven days building its nest on the rocks close by the shore, and the second seven days hatching its young, protected all the time by the fine weather.

But the respite was brief. The halcyon days, rather like our own, were soon over. As Virgil put it, describing the return of winter, and the rising wind: 'The flimsy gossamer now flits no more/Nor halkyons bask on the short sunny shore.'

Shawondasee's Summer

Sometimes, in the twilight of the year, the weather gives us one last chance, a so-called 'Indian summer'. The term is American. It applies to a period of quiet and unseasonably warm weather any time between September and early December.

The story goes that the American Indians spotted that such a spell regularly occurs on the Great Plains in late autumn or early winter. Moreover, since a fine spell of this kind is generally associated with an anticyclone, resulting in very little vertical motion in the atmosphere, visibility is often reduced with haze and smoke. This aspect of the Indian Summer is also reflected in the folklore in which it has its origins.

According to one Indian legend the god Nanahbozhoo always sleeps during the winter, but before retiring in the autumn he fills and lights his great pipe and smokes for several days. Good weather is always assured for this ceremony, and it is smoke from the pipe of Nanahbozhoo which causes the haziness during the fine spell.

Longfellow has another story in 'The Song of Hiawatha'. According to his version, the Indian Summer is brought by *Shawondasee*, the personification of the South Wind, a languid, warm, and easy-going spirit:

Shawondasee, fat and lazy,
Had his dwelling far to southward
In the drowsy dreamy sunshine,
In the never ending summer.

Shawondasee was benign. He sent the migratory birds northward in the springtime, and provided the right conditions for fertile crops and rich hunting. And before the real onset of winter each year ...

From his pipe the smoke ascending,
Filled the sky with haze and vapour
Filled the air with dreamy softness
Gave a twinkle to the water
Touched the rugged hills with smoothness
Brought the tender Indian Summer
To the melancholy North land
In the dreamy Moon of Snow-shoes.

Index